HUMAN REPLAY

A Theory of the Evolution of Media

Paul Levinson

Connected Editions

Copyright © 1979 Paul Levinson

All rights reserved

No part of this book may be reproduced, or stored in a retrieval system, or transmitted in any form or by any means, electronic, mechanical, photocopying, recording, or otherwise, without express written permission of Paul Levinson.

ISBN-13: 978-1-56178-062-4

Printed in the United States of America

CONTENTS

Title Page
Copyright
Notes from the Author, 2017, 2022 — 1
HUMAN REPLAY: — 3
Sponsoring Committee — 5
Abstract — 6
PREFACE — 10
CHAPTER 1 — 14
INTRODUCTION: TWO DIFFERENT CREATIONS? — 15
The Anthropotropic Model: Reentering the Garden with Knowledge — 22
Leads in the Literature — 27
Testimony of the Media — 30
The Falsificationist Approach — 34
Limitations: Where the Theory Needn't Go — 37
PART I — 46
ROOTS AND THREADS OF ANTHROPOTROPIC THEORY — 47
CHAPTER 2 — 48
TECHNOLOGY AS EXTENSION — 49
Sigmund Freud: The Prosthetic God — 50
R. Buckminster Fuller: Eyes Equal Eyeglasses — 56

Edward Hall: Spiders' Webs and TV Networks — A Biological Purpose for Extension — 60

Marshall McLuhan (I): Humans As Technological Effects — 63

Marshall McLuhan (II): Oral, Written, and Electronic — An Anthropotropic-Like Schema — 71

CHAPTER 3 — 78

INTIMATIONS OF MEDIA EVOLUTION — 79

Peter Medawar: Agenda for a Theory of Technological Evolution — 80

Lamarck, Darwin, and Teilhard: Models of Evolution — 83

Lewis Mumford: The Lost Discovery — 90

Harold Innis: A Dialectic of Media — 95

CHAPTER 4 — 106

ANALOGUES AND PARTICULARS — 107

Noam Chomsky: Deep Structures and Transformations — 109

Claude Levi-Strauss: Bi-Polar Mentalities — 120

Karl Popper: A Darwinian Evolution of Ideas — 123

Norbert Wiener and Cybernetics: The "Why" of Anthropotropic Evolution — 137

Brains, Computers, and Minds — 143

The Medium That Became the World: Teilhard's Noosphere — 146

Siegfried Kracauer: Film As a Happy Ending — 153

André Bazin: The Anthropotropic Intention — 157

PART II — 168

THE ANTHROPOTROPIC EMERGENCE — 169

CHAPTER 5 — 170

ORIGINS AND ASCENT OF THE SPECIES — 171

Hieroglyphics and Phonetic Writing: The Primitive — 173

Syndrome

Print: At the Turning Point — 178

Telegraph and Photography: The Two Great Branches — 183

Telegraph: Retrieval of the Pre-Technological Process — 185

Photography: Retrieval of the Pre-Technological Content — 191

Motion Photography: Beyond the Landscapes of Theater — 197

Phonograph: "Photographing" the Sounds of Life — 203

"Talkies": Reunion of Sight and Sound — 207

Color: Less than Life, More than Life — 209

Electronic Co-option of the Photographic Branch: Reunion of Process and Content — 212

Telephone and Radio: Interaction vs. Access — 215

Television in the Vanguard: Life-Size Screens, Home Video Recording, Holography, Fiber Optics — 218

CHAPTER 6 — 229

Media in Human Ecological Niches — 230

Precision vs. Scope of Approximation — 237

The Principle of "Net Gain" — 239

Replication vs. Extension — 241

Co-evolution and Convergence — 243

Persistence of Abstraction — 247

CHAPTER 7 — 272

Mobility and Access: Wires, Airwaves, and Light Beams — 273

Observational Terminus: Holography and the Retrieval of the Third Dimension — 277

The Aging Heir Apparent: Videophone — 281

Interactive Terminus: The Reunion of Talking and Walking — 287

CHAPTER 8	291
SUMMARY	292
EPILOG	299
CHAPTER 9	300
THE CENTER BEHELD!	301
Appropriate Monsters	302
The Dream and the Act: The Irreducible Difference	304
The Human Option	307
BIBLIOGRAPHY	312
Books	313
Articles and Miscellaneous Printed Material	320
Non-Book, Non-Periodical Media	328
GLOSSARY	330
terms used in this dissertation	331
About the Author	335
The following books by Paul Levinson available in print and Kindle:	337

NOTES FROM THE AUTHOR, 2017, 2022

The following is my PhD dissertation, almost exactly as it was submitted and defended at New York University in 1978. I say "almost," because I moved the footnotes at the bottom of pages to the bottom of paragraphs to accommodate easier Kindle formatting, and corrected one typo (a date) which survived countless readings by me and others back then. (That typo led to my view that there is an irreducible level of entropy in the process of writing, editing, and publication.) The "About the Author" section at the end of the dissertation is also current.

You'll find here, in addition to creation and construction of my "anthropotropic" theory of media evolution, extensive discussion of Marshall McLuhan – who, I as thought back then, is every bit as important a guide to media today, some 40 years later. You'll also find mentions of thinkers who would go on to become important media theorists, including Joshua Meyrowitz and Bob Logan. Among the many predictions in this dissertation, I guess the one I'm most proud of is the arrival of the iPhone in 2007, fulfilling what I wrote here in 1978 that "the wireless, portable evolution of media should continue to the point of providing any individual with access to all the information of the planet, from any place on the planet, indoors and outdoors, and of course even beyond the planet itself as communication extends into the solar system and cosmos beyond." Although I wrote and successfully defended this dissertation in 1978, it carries the date of 1979 because the defense took place in the Fall, and New York University therefore considers the dissertation completed for my February 1979 graduation.

Tina Vozick was my wife in 1978, she still is — I thanked her then and thank her again for suffering through my writing of this dissertation and now for inspiring me for everything since. Our children, Simon and Molly, and our grandson, Mikey, have all inspired me from the days they were born.

— Paul Levinson, New York, September 2017

Our second grandson, Julian, has also inspired me from the moment of his birth.

And I thank Andrey Mir for his careful reading of this dissertation in the past year, finding too many typos in the scanning of the original document (now corrected).

— Paul Levinson, New York, March 2022

HUMAN REPLAY:

A THEORY OF THE EVOLUTION OF MEDIA

PAUL LEVINSON

Submitted in partial fulfillment of the requirements for the degree of Doctor of Philosophy in the School of Education, Health, Nursing, and Arts Professions of New York University.

1979

I hereby guarantee that no part of the dissertation which I have submitted for publication has been heretofore published and (or) copyrighted in the United States of America, except in the case of passages quoted from other published sources; that I am the sole author and proprietor of said dissertation; that the dissertation contains no matter which, if published, will be libelous or otherwise injurious, or infringe in any way the copyright of any other party; and that I will defend, indemnify and hold harmless New York University against all suits and proceedings which may be brought and against all claims which may be made against New York University by reason of the publication of said dissertation.

SPONSORING COMMITTEE

Professor Neil Postman, Chairman

Professor Christine Nystrom

Professor Ronald Todd

ABSTRACT

HUMAN REPLAY:

A THEORY OF THE EVOLUTION OF MEDIA

PAUL LEVINSON

Submitted in partial fulfillment of the requirements for the degree of Doctor of Philosophy in the School of Education, Health, Nursing, and Arts Professions of New York University.

1978

Although much has been written about how artificial media influence and even direct our lives, few people have inquired as to how media themselves change, and what role humans may play in directing that change. Several recent developments, indeed, suggest a definite pattern to media change, a pattern in which human direction figures very prominently. Television, for example, has changed from black-and-white to color; the telephone has all but supplanted the telegraph; music recordings are increasingly played through multiple rather than single speakers. When one considers that h u mans in the natural or "pre-technological" state see in colors rather than black-and-white, speak in voices rather than Morse code, usually hear sounds emanating from a variety of sources rather than a single source, the pattern of media change becomes fairly clear: media are evolving, not to more artificial forms, but to reproduction of human or "pre-technological" forms of communication.

The study attempts to develop this observation into a general theory of media evolution, that describes, explains, and predicts the evolution of media. The methods entail an examination of previous thinking about the nature and evolution of media, and an examination of the development of media themselves.

Although few theorists have examined the relationship of media and human communication systems in depth, partial recognition of the increasingly human pattern of technological communication has been made by a variety of observers, in a variety of times and places. Victorian novelist Samuel Butler was one of the first to suggest that technologies are, in effect, artificial limbs or organs,' which function as extensions of the human system. Psychologist Sigmund Freud pointed out that humans use their technological extensions to become "prosthetic" gods, in fulfillment of fundamental human desires. Marshall McLuhan, who has perhaps probed media more deeply than anyone to date, actually anticipates the present thesis in his suggestion that electronic media are retrieving the communication environments of primitive, tribal societies.

The evidence of media evolution appears to confirm the thesis of the study. Photography, for example, even in its initially motionless, speechless, colorless state, provided a more literal replication of the real world than both the printed word and all but the most true-to-life, colored painting. And with the addition of motion, sound, and color, and the subsequent attainment of immediacy through television, and even the third dimension through holography, the technological perception of the world through lenses and microphones has become ever more like the human perception of the world through eyes and ears.

From these and similar examples, the study abstracts an overriding "principle" of media evolution: namely, that media survive based on how well they replicate a human mode of communication. Poor replicators, like telegraph and silent movies, tend to disappear; better replicators, such as. telephone and "talkies," survive until a more complete replicator is invented. In effect, media seem to evolve in a Darwinian-like process of natural selection, with humans acting as the selectors. Such a model is supported by the work of philosopher Karl Popper, who suggests that all products of human mentalities evolve in a Darwinian-like pattern.

From the application of this principle and its corollaries, predictions of future media are made — the most striking of which suggests that humans will someday be instantaneously transported across vast distances, as the acts of transportation and communication are reunited.

Should such a development be feared? Probably not, for the study concludes that we need have no less control over our most advanced technologies than we do over our primitive technologies like the window, which is controlled by merely pulling down the window shade.

"In short, by the use of our hands, we bring into being within the realm of Nature, a second nature for ourselves."

Cicero

On the Nature of the Gods

Book II, p. 60

circa 45 BC

Translation appearing in Benjamin Farrington , *The Philosophy of Francis Bacon* (Liverpool Liverpool Univ. Press, 1964), p. 28

PREFACE

The observation that communications media are the vital life-lines of our world has become almost commonplace. Arthur C. Clarke perhaps put it more ingeniously when he noted recently that humans can function longer without food and water than without information.[1] And indeed, as the most social of all animals, we are thoroughly and utterly dependent upon knowing what is going on. Our lives course through the wires, airways, print and photo-chemical compounds of communications technologies as surely as our lives course through the nervous systems of our bodies.

1. "Communications in the Second Century of the Telephone," in The Telephone's First Century — and Beyond, with a Preface and Afterword by John D. deButts, and an Introduction by Thomas E. Bolger (New York: Crowell, 1977), p.86.

Understandably, we are concerned with what these communication technologies may be "doing" to us. These media are, after all, not of our bodies in the way that ganglia and neurons are. Much has been written about how these interlopers "influence" us — how media shape and direct our feelings, thoughts, societies, and, inevitably, our lives.

Yet largely absent from most of these analyses is any inquiry into what drives the media that drive us. Few observers of media seem to notice that media themselves are changing; even fewer think to ask if there is some pattern to that change.

The ensuing study will inquire into how media change. Moreover, it will look for a discernible pattern in that change, and thereby in effect assume that media do more than merely change. For to change in a pattern that is comprehensible to rational observation — to change in response to knowable forces, whose influences may be charted, perhaps reduced to principles and even predicted — is to change in an organized way that is called evolution."

The study will begin, moreover, with a good idea as to the nature of that pattern of media change or evolution. The idea stems from several simple observations of media change. We now watch color television where we used to watch black-and-white? we generally talk on the telephone rather than send telegrams; the single, tin"horn" of the old Morning Glory Victrola has blossomed into a multi-phonic music presentation of such range and subtlety that it more than rivals the acoustics of the concert hall.

What have these changes in common? In a world without technology, we obviously see in color, speak with voices rather than Morse code, and hear sounds emanating and bouncing off multiple sources. It seems as if all these developments in media have in some essential way made the media less technological or media-like — that, somehow, these new media are less artificial than their predecessors, recapturing a fuller texture of the real world. It may perhaps be said that these new media "replay" the world of colors, voices, and multi-dimensional sounds with which human communication began.

This study will attempt to expand these observations into a general theory of media evolution. Two approaches will be used: Part I will examine existing literature on media and related areas for previously recognized clues as to the evolution of media along "human" lines. These will be found, surprisingly, in a wide array of places ranging from Freud's psychoanalytic perspective to Teilhard de Chardin's religious speculation. They will also he found, not so surprisingly, in the works of the great pioneering observers of media and technology like Marshall McLuhan, Harold Innis, and Lewis Mumford — though here an over-concern with the impact rather than the origin of media, alluded to above, has often distracted these discoverers from the significance of their own discoveries.

Being in a position to benefit, however, from these previous insights, Part II of the study will present a full chronicle of the actual evolution of media, from the beginnings in hieroglyphics to present-day developments in three-dimensional holography. Consistent with the thesis suggested above, this chronicle will attempt to show how media, with increasing success, have managed to reproduce living or human patterns of communication — how they have evolved, in effect, in response to the human communication environment of colors, voices, depth perception, and so on. Such an account should suggest principles or mechanisms that govern the survival and evolution of media; and these, in turn, should help better describe and explain media developments of the past, and perhaps even predict, in a general way, developments of the future.

From these two approaches — examining the previous thinking about media, and examining the media themselves — a general theory that describes, explains, and predicts the evolution of media should emerge. The theory will be rational, scientific, and new. It will be rational because its assumptions and arguments will be exposed rather than implicit or hidden, and open at all points to logical refutation by a more reasonable explanation. The theory's science will reside in the rationality of its method, and the "falsifiability," in Sir Karl Popper's terminology, of its conclusions — thus, descriptions of the past and present will be amenable to alteration or contradiction by overlooked evidence, and predictions of the future will of course stand or fall in their own time. (Though here it must be cautioned that if strict quantification and control are to be regarded as essential to scientific inquiry, then this theory will fall far short of the mark, as perhaps would any attempt to theorize about human affairs.) The newness of the theory will, in the long run, be its least important attribute, and reflects only the author's view that nothing quite like it has thus far been attempted.

All of the foregoing, to be sure, in no way assures that the theory will be correct. But if this theory is even partially right — that is, if it is even partially less incorrect than its predecessors — then a "new" element will be injected into the discussion of media impact: the human impact, or the human shaping of media. The only thing surprising about this perspective is that it has been ignored or taken for granted for so long: for just as our thoughts and words are indisputably products of our human mentalities, so too are our media.

***

There are many friends, colleagues, teachers, and students who have helped make this work possible. There is only one person, however, without whose encouragement, suggestions, and criticism, this work would have been impossible: my wife, Tina Vozick.

CHAPTER 1

INTRODUCTION: TWO DIFFERENT CREATIONS?

Writing in *The Cosmic Connection*, Carl Sagan offers the following account of the origin and essence of life:

> Eventually a molecule was formed that had a remarkable capability. It was able to produce, out of the molecular building blocks of the surrounding waters, a fairly accurate copy of itself. In such a molecular system there is a set of instructions, a molecular code, containing the sequence of building blocks from which the larger molecule is constructed. When, by accident, there is a change in the sequence, the copy is likewise changed. Such a molecular system — capable of replication, mutation, and replication of its mutations — can be called 'alive.' It is a collection of molecules that can evolve by natural selection. Those molecules able to replicate faster, or to reprocess building blocks from their surroundings into a more useful variety, reproduced more efficiently than their competitors — and eventually dominated.

[1]

1. (New York: Dell, 1973), pp. 3-4

The postscript was provided by Sagan on a *Tonight Show* appearance:

> And then came a creature whose genetic material was in no way different from the self-replicating molecular collectives of any of the other organisms on his planet. But he was able to ponder the mystery of his origins . . . . He was the matter of the cosmos contemplating itself. . . . [2]

2. As reported by Stuart Baur, "Kneedeep in the Cosmic Overwhelm with Carl Sagan," *New York Magazine*. 1 September 1975, p. 30

This capacity to replicate has been celebrated by numerous theorists as the fundamental defining characteristic of life[3]; yet what pertains to the blind process of genetics is often thought to be inapplicable to the conscious expression of life in human intelligence; many of the artifacts and technologies created by human intelligence bear little outward resemblance to their creators, and indeed often seem to impede and disrupt rather than perpetuate the human condition. Imperfect and inhuman, cast as the cliched Frankenstein monster, technology is condemned for severing the human race from its past, for luring us with mirage and facade into a shadow world of non-reality and even non-existence. From such a perspective our species appears to be the victim of a grotesque conflict; integrated, by virtue of our genes, into the grand replicative design of life, yet increasingly removed from that design by and with the compliments of our culture; the product, as it were, of two creations.

3. See, for example, Richard Dawkins, *The Selfish Gene* (New York; Oxford University Press, 1976).

Perhaps at once the most eloquent and extreme presentation of that view has been made by Jacques Ellul, who sees the human attempt to organize the environment, first through magic and then increasingly scientific procedures, as driving an ever-widening wedge between our humanity and our civilization, culminating in today's "Technological Society":

Man was made to do his daily work with his muscles; but see him now, like a fly on flypaper, seated for eight hours, motionless at a desk. . . . The human being was made to breathe the good air of nature, but what he breathes is an obscure compound of acids and coal-tars. He was created for a living environment, but he dwells in a lunar world of stone, cement, asphalt, glass, cast iron, and steel. The trees wilt and blanch among sterile and blind stone facades. Cats and dogs disappear little by little from the city, going the way of the horse. Only rats and men remain to populate a dead world. Man was created to have room to move about in, to gaze into far distances, to live in rooms which, even when they were tiny, opened out on fields. See him now, enclosed by the rules and architectural necessities imposed by over-population in a twelve-by-twelve closet opening out on an anonymous world of city streets. [4]

4. Jacques Ellul, The Technological Society, trans. John Wilkinson (original ed., n.p., 1954; trans. ed., New York: Knopf, 1964), p. 321.

And chief among the technological villains, Ellul adds, are communications media like movies and television, which lead us "straight into an artificial paradise," to do their worst by providing "an absolute distraction, a total obliviousness" to the very real malaise that technology engenders. [5]

5. Ibid., pp. 377-380.

And indeed, when one considers the housewife staring for hours at black-and-white two-dimensional flickers on a tiny screen, the adolescent making verbal love to a faceless voice on the telephone, and even, or perhaps most of all, the scholar and seeker of truth pouring over squiggles on a page that have only the most arbitrary connection to the real world of sounds and images—when one considers these various testimonies of the media, they do seem to plead guilty to Ellul's charge of gross subversion of reality.

For doesn't reality come in colors and not in black-and-white? And isn't the world, in the absence of communications media, constructed along three rather than two dimensions? And where, other than in print, are the lines, angles, and curves that form "tree" taken for that leafy green growth with a rough brown trunk in the backyard? Nowhere in nature, only in media.

But what if media were able to replicate images in color, and events in three dimensions? What if you could communicate with someone over vast distances almost instantaneously and perceive the face along with the voice? What if music recording were more than a thin, reedy echo of the original; what if it were able to retrieve the rich texture of tonal nuances so that it was virtually indistinguishable from the original? When Ellul wrote *The Technological Society* in 1954, these media were all but nonexistent; now they are either in widespread use or on the verge of entering our culture.

The triumph of color television is documented by the recent report that more than 75% of American homes now enjoy such service;[6] still and moving holograms, which present three-dimensional images viewable from numerous angles, are now on display in several cities in the U.S. [7] The possibility of conversing with pictures — first introduced with Bell Telephone's "Picturephone," and more recently through interactive or two-way cable television systems — is now an actuality in cities such as Chicago and Reading, PA. [8] In the recording and reproduction of sound, the early monaural techniques have multiplied into stereo, quadraphonic, and systems of even subtler differentiation, which provide a "surround" of sound comparable to that heard in a stroll down the street or a walk in a forest. [9] And in the realm of intellectual discourse, print's role as sole conduit has been readily supplemented by more life-like sounds and images, as shown by the widespread educational usage of such television programs as "The Age of Uncertainty," "The Ascent of Man," and "The Adams Chronicles," to mention but a few.[10]

6. A. C. Nielson Company, *Nielson Television 1977* (Northbrook, IL: A.C. Nielson, 1977), pp. 4-5.

7. Steve Ditlea, "What is a Hologram? How Does It Float in Midair . . . And Is It An Art?" *Ms.*, December 1976, pp. 34-39. See George W. Stroke, An Introduction to Coherent Optics and Holography, 2nd ed. (New York: Academic Press, 1969) for a discussion of the historical development and physical principles of holography.

8. See Arthur Gregor, Bell Laboratories (New York: Charles Scribner's, 1972), pp. 81-91, for a review of Bell Telephone's work on the "Picturephone"; see NYU-Reading Consortium, Final Report: Berks Community Television (New York: NYU-Reading Consortium, forthcoming) for details on the use of interactive cable television in Reading, PA.

9. See Michael Zilkha, "Future Hock," New York. 19 April 1976. pp. 6 6 -6 8 , for a brief description of some of the latest new audio and audio-visual devices now available to the public.

10. For example, Pat Thaler and Sonya Shapiro reported in "New Routes to a College Degree," *New York*, 29 August 1977, p. 40, that several colleges planned to offer academic credit for viewing the regular broadcasts of "The Age of Uncertainty" on WNET-TV in the Fall of 1977. A telephone

interview with Talia Gross of WNET-TV in New York City on May 2, 1978 confirmed that six colleges in the New York metropolitan area did in fact offer such credit.

Ellul would perhaps argue that in achieving such proximity to the real world, the new media provide an even more tempting distraction from that real world, thus working to the even greater detriment of the human condition. Yet as the technological threat to reality becomes more like the reality itself, the threat in some fundamental sense surely becomes less of a threat. And in any case, the new developments in media seem to contradict what might be expected of media evolution on the basis of Ellul's observations: for rather than producing communication environments that are increasingly artificial or different than the environment of the real world, media seem to be producing environments that are less artificial or more like human, pre-technological situations. Color and three-dimensional media are simply less artificial and inhuman than the black-and-white and two-dimensional devices that preceded them.

And this "humanizing" phenomenon appears to be spreading. In the area of remedial media, a recent TV newscast describes a new "living" contact lens, made of plastic so soft and permeable that it "breathes" and can function as an almost permanent accessory of the human eye.[11] And several months later, the same news program reports that "micro-electronics are getting closer to the kind of miniature electronic devices common in nature, [12] such as our nervous systems," thus indicating yet an additional element in the correspondence between computers and brains first noticed nearly thirty years ago by Norbert Wiener. [13]

11. Frank Field, WNBC-TV, "News Center Four," 1 November 1976, 6:25 PM.

12. Robert Potts, WNBC-TV, "News Center Four," 3 March 1977, 6:50 PM.

13. See "Norbert Wiener and Cybernetcis" in Chapter 4 below for a discussion of Wiener's work in communications, and its relevance to the present study. See Chapter 7, n. 1 below for more on micro-computers and electronics.

At the same time, moreover, a brief scan of history reveals this tendency of technological media to increasingly replicate the living, real world as having been operating, albeit quietly, for quite some time. The invention of photography in the 1820s was of course itself perhaps the single most vivid leap in the technological ability to recapture the real world; and with each subsequent innovation in photographic process, from still to motion, to synchronized sight-and-sound and color, to the instantaneous transmission of sight and sound that is technically no longer photography at all but electronic video, the apparatus of replication has been endowed with a capacity to retrieve more and more of the real world. Less complete, but equally indicative, has been the supplantation of the telegraph with the telephone, allowing for a more human, though still less than totally human, two-way communication via voices rather than dots and dashes.

THE ANTHROPOTROPIC MODEL: REENTERING THE GARDEN WITH KNOWLEDGE

O n the basis of these scattered and preliminary observations, it is possible to discern a systematic pattern in the development of technological media: namely, that as technological communications media evolve, they tend to increasingly replicate the pre-technological or human communication environments of the real world.

The significance of this evolution becomes clearer when the difference between technological and non-technological communications is defined. Technological communication, as in print, telegraphy, and photography, occurs only through the intervention of a mechanical, electronic, or otherwise artificial contrivance; non- or pre-technological communication,on the other hand, is accomplished without artificial intervention, as in the case of "face-to-face" communication or any unaided observation of surroundings. (Note that "non"- or "pre"-technological situations exist both historically, or before the introduction of particular technologies, and currently, as when we converse in person or gaze at the stars.)[14]

14. With regard to the second point, it might be argued that once a technology has been introduced, purely "non"-technological activity is impossible, since our actions and perceptions in the absence of the technology are conditioned by our experiences with the technology. Examples of this sort are to be found when we talk to people without looking at them, as if we were on the telephone, or when we walk down the street with the sensation that we're in a movie. The limitations of such technological effects, however, lie in the fact that a century of telephone has yet to make us blind, and nearly as much experience with flat film has yet to delete the perception of depth from the real world. Fundamental patterns of pre-technological communication, in other words, seem unchanged by modern technologies. But see also the ends of Chapters 2 and 6 below. (See also Glossary at conclusion of study for definitions of terms used in the study.)

The evidence described above suggests that as technological media become more sophisticated, they produce or allow for communication that increasingly resembles non-or pre-technological communications. Motion photography and video-phone are more akin to face-to-face communication than are still photography and telephone.

But if technology ultimately only duplicates the non-technological, why were technologies invented in the first place? Because the advanced-technological and the pre-technological still differ in one all important respect. As numerous media theorists have pointed out (see Chapter 2 for a fuller discussion), technological communication has attempted to overcome the limitations of space and time that are a part of pre-technological environments. Thus, telephone extends the ear across distances unbridgeable in the pre-technological situation, as photography extends the eye across formerly unviewable time. But to transcend the real world, primitive technology must sacrifice the real world: telephone extends the ear at the expense of the eye, and the first photographs extended sight at the expense of motion, sound, and color. But of what value are extensions that enrich on the one hand and impoverish on the other? It becomes the function of advancing technology, then, to both extend across time and space and recoup elements of the real world lost in prior extensions. Thus, video-phone and motion pictures.

A revised three-stage model for the evolution of technological media thus emerges: (A) In the beginning, all communication is non-technological or face-to-face. All elements which characterize real-world perception, such as color and motion, are present. Biological limitations on the ability to communicate across space and time are also present. (B) Technologies are devised to overcome the biological constrictions on space and time communication. But to overcome these limitations, early technologies must jettison many desirable components of the real world environment, such as color and motion perception. (C) As technologies are made more sophisticated, they attempt to regain the elements of face-to-face communication lost by earlier technologies, as well as maintain (and in some cases improve) the extension across space and time. Advanced technology thus combines the extension of Stage B with the reality of Stage A — allowing us to have the cake of our extensions and eat it too. Or, if technology is the apple of knowledge that at first extended us out of the natural Garden of Eden, it turns out in the end to h e a remarkable apple that permits us re-entry.

But because Stage C is a still emergent process, it is not nearly as much in evidence as the vast artificial legacy of Stage B. Indeed, the condemnation of technology by Ellul and other critics would seem the only reasonable outcome of investigations overwhelmed by the evidence of Stage B. Yet to the extent that recent developments in communication technologies are not "flukes," but indicative of a major evolution, it would seem that Ellul's view of technology as at odds with life, is, for communication technologies at least, somewhat premature. For as technology extends nature rather than circumvents it — as it attempts to overcome biological limits by complete re-creation instead of incomplete substitution — the opportunities only increase, to paraphrase Carl Sagan, for the matter of the cosmos to contemplate itself.

It will be the purpose of the present study to examine the often fragmentary and as yet tentative evidence for the emergence of Stage C communication technologies, and to fashion from it a general theory of media evolution which puts the past development of communications media in focus, accounts for the evidence of the present, and leads to predictions about the future. Since the label "Stage C" is a hit inelegant and non-descriptive, the equally inelegant hut perhaps more explanatory term "anthropotropic" will he coined to describe the theory entailing the view that media are evolving towards increasing replication of "face-to-face" or human communication environments, though without the biological constraints of space and time ("anthropo"=human; "tropic"=an affinity for).

It is obvious that such a proposition would have been difficult to formulate even 25 years ago, and for the same reason much easier to substantiate— or refute— 25 years hence. Indeed, many fine details of the theory will no doubt have to wait for the passage of time, and continued development of communications media (assuming these developments don't render the theory untenable). Nevertheless, this inquiry will proceed on the basis of there being sufficient evidence at hand with which to at least initially construct such a theory, and will seek such evidence from two distinct, though obviously overlapping, sources:

first, the observations and conclusions of those who have already given much thought to communication technologies and related human activities; and next, the actual historical development and current status of the communications media themselves.

LEADS IN THE LITERATURE

Considering others' observations of technologies before considering the technologies themselves may seem a bit of a transposition, but is consistent with the view that direct observation is not a particularly productive activity, unless one has some idea, at least, of what to look for. Moreover, since even Athena, the goddess of wisdom, sprung fully grown from Zeus' head only after Zeus had ingested his consort Metis (an older source of wisdom) and suffered from a consequent headache, [15]it seems only reasonable that a mere theory derive in the first place from the wisdom (and headaches) of previous thinkers on the subject. Thus, the lessons of previous inquiries into media and related fields will serve as the "ground," to use a Gestalt term recently revived for media study by Marshall McLuhan, [16] upon which the present theory will be erected.

But prior perceptions of media will serve as more than a passive backdrop or source of orientation for the present study. For, as Thomas Kuhn has pointed out in *The Structure of Scientific Revolutions*, [17] a successful new theory must not only explain that which has been previously left unexplained, but account for all that has previously been explained: a new theory, in other words, should generate a net gain in understanding, by taking into account not only the lapses, but the insights, of previous theories. In the study of media, however, such a process is both simplified and complicated by the fact that there have been few theories that have attempted to fully address the evolution of technological media — indeed, if, as in Kuhn's parlance, a "pre-paradigmatic" state of inquiry is one in which there are no clearly commanding models or theories, than the area under present inquiry must be judged as a "pre-pre-paradigmatic" state, for here there are by and large no fully-developed theories at all.

15. H. J. Rose, *A Handbook of Greek Mythology* (New York: Dutton, 1959), p. 50.

16. See, for example, Marshall McLuhan, Kathryn Hutchon, and Eric McLuhan, *City As Classroom* (Agincourt, Ontario: Book Society of Canada, 1977), pp. 8-31.

17. 2nd ed. (Chicago: Univ. of Chicago Press, 1970).

Nonetheless, media and technology have been the subject of a considerably rich array of insights and understanding, both in the asides of theories directed at other aspects of the human condition, and in the unsystematic though lively scrutiny of the media themselves. Many of these observations, moreover, may seem to contradict the thrust of the present thesis, and must therefore receive special attention. Thus, McLuhan's central observation, for example, that media tend to distort or unhinge human sensory ratios — for which he offers much persuasive evidence and argument — must be accounted for by a theory which suggests that media are evolving towards greater consonance with human systems.

Yet a third benefit to be accrued from the raking through of previous work in media is the co-option of applicable models and even jargon. The Shannon-Weaver "mathematical" model of communication, for example, speaks of the "encoding" and "decoding" of information that characterizes its transmission through all media. [18] Presented in Shannon-Weaver terms, anthropotropic evolution might be described as a movement of media from minimum encoding and maximum decoding to maximum encoding and minimum decoding — that is, as media become more advanced, they transmit information that requires less and less decoding on the part of the perceiver (and commensurately more encoding by the producer). Thus, television pictures are more immediately comprehensible — require less decoding — than written words (TV, unlike reading, need not be "taught"); although television production is a considerably more complex task, technologically, than writing. Such a trend towards media presentations that require successively less decoding can be traced in evolutionary sequence from print, which must be decoded into the full human sensorium of sight, sound, etc.; to radio, which in offering sound needs decoding only into sight, etc.; to television, whose color, sight-and-sound information requires decoding only into the third visual dimension; and so forth. Such borrowing of constructs has of course long been one of the staples of scientific inquiry (indeed, as will be seen in ensuing chapters, much of the perspective of the present theory derives from Darwinian and other models of organic evolution).

18. Warren Weaver, "The Mathematics of Communication," in *Communication and Culture*, ed. Alfred G. Smith (New York: Holt, Rinehart, and Winston, 1966), pp. 13-24

TESTIMONY OF THE MEDIA

Having thus culled the existing literature for leads, danger signals, and metaphors, the present study will then proceed to examine the actual incidence of anthropotropic media evolution. This "empirical" part of the study will utilize several approaches.

To begin with, the entire evolution of media, from hieroglyphics to holography, will he examined for evidence of an anthropotropic movement. And exactly what will constitute such evidence? In general, the growing preponderance, throughout history, of media that replicate more of the pre-technological human communication environment than their predecessors (such as the example of photographic development cited earlier). Certain exceptions to this pattern, however, may he expected to arise since advanced media attempt not only to replicate the real world, hut to continue and even improve the extension across time and space of more primitive media: thus a medium may on occasion he succeeded hy one which does a slightly worse job of replication, while a vastly superior job of extension (such was indeed the case in the change from hieroglyphics to phonetics, and print to telegraph — see Chapter 5 below for details). Such cases will of course merit special attention and explanation. In the long wash of history, however, the poorer replicators would be expected to gradually drop out in favor of their more life-like cousins.

Evidence of a more psychological nature will also be admissible. Thomas Edison, for example, is said to have intended his motion picture invention to include synchronized sound (indeed, Edison reportedly developed his motion picture process in the first place as a pictorial accompaniment to his previously invented phonograph) ; [19] his actual failure to perfect such a workable talking picture process, then, seems more reasonably attributed to a lack of technological know-how than a lack of desire, and should thus be acceptable as evidence of the human attempt to fashion media in increasing correspondence to human communication environments. Indeed, as will be seen in Chapter 5 below, such anthropotropic intentions have almost always been realized by subsequent technologies — not only in the case of "talkies," but in nearly every other instance — and thus serve as important indicators of technological evolution.

19. Matthew Josephson, *Edison* (New York; McGraw-Hill, 1959), p. 385; see also Gerald Mast, *A Short History of the Movies* (New York; Pegasus, 1971), pp. 25-26

After thus surveying the historical development of media with an eye for evidence of the present thesis, the study will then attempt to abstract from such evidence a series of principles or "rules" that seem to govern the course of media evolution towards recreation of human environments. The principles that regulate the evolution of any system — organic, media, or otherwise — in effect answer the question of what a member of the system must do, or be, to survive. In general Darwinian (or, for that matter, Lamarckian) terms, an organism survives by accommodating the requirements of its environment better than any of its rivals; in general anthropotropic terms, a medium should survive by approximating pre-technological communication better than any of its competitors — that is, media that closely replicate the pre-technological world should have a better chance for survival than media that don't. But, as in the biological realm, the practical operation of such a general principle in media evolution has been more complicated: no medium has yet, for example, approximated the entire pre-technological communication environment, so questions arise as to which aspects of pre-technological communication figure more importantly in a medium's survival, with what precision must a medium replicate in order to survive, and so forth.

The answers to such questions — and the principles of media evolution they will suggest — lie in the survival, and non-survival, records of specific media observed in the historical survey. Of greatest value here will be those cases of media that seem to run contrary to anthropotropic expectations — of media surviving when they "shouldn't," and media not surviving when they "should." A good example of the first type is the survival of the telegraph, which, as a transmitter of electronic beeps, was in all ways a poorer replicator of reality than the telephone, and thus should have vanished with the advent of the talking phone. Explaining the telegraph's survival by noticing that it, unlike the telephone (until recently), provides ready permanent transcriptions of all its transmissions, suggests that a medium's longevity depends not only upon its ability to replicate the pre-technological environment but upon its capacity to extend across time (and by implication, space), and that the principle of pre-technological approximation must be accordingly refined. An example of the second type arises in the rapid demise of the "3-D" movie experiment of the 1950s, which, in providing the third visual dimension of depth, should have attained a survival edge over television and motion pictures, its two-dimensional predecessors. A careful look at 3-D, however, reveals that its third dimensional effect was perceivable only to those viewers who wore special glasses and gazed straight ahead — that is, the natural dividend of the third dimension was collectable only under unnatural circumstances — which suggests that for a medium to survive, it must not only replicate some aspect of the pre-technological environment, but do so in a way that doesn't distort other elements of pre-technological communication.

THE FALSIFICATIONIST APPROACH

Discovery and refinement of principles through confrontation of seemingly contradicting cases is consistent with the consideration of opposing theoretical frameworks discussed earlier in this chapter, and is part of an approach to scientific and philosophic inquiry first introduced by philosopher Karl Popper, which will serve as a main posture for the present inquiry. A practical example of this confrontational or "falsificationist" approach, as it is known, has been provided by Popper's student Bryan Magee [20]: Suppose, Magee asks, we are interested in investigating the boiling point of water, which we "know" or expect to be 100 degrees Centigrade. By far the most interesting areas of investigation are provided by observations that tend to contradict our initial expectations about water — such as water not boiling at 100 degrees C in closed vessels, or at high altitudes. In accounting for these contradictions to our initial hypothesis, we come to understand more about the properties of water — that the boiling point of water is not only a function of temperature, but of atmospheric pressure, etc. "We are challenged to produce a hypothesis," Magee explains, "altogether richer than our original, simple statement . . . a hypothesis which would explain why our first hypothesis worked, up to the point it did, but then broke down at that point? and also enable us to account for the new situation as well . . . . our revised theory, whether true or false, would thus tell us more about the world than we yet knew."[21]

20. Bryan Magee, *Karl Popper* (New York: Viking, 1973), pp. 16-18.

21. Ibid., pp. 17-18.

The brief explanation, attempted above, of the apparent exceptions to the anthropotropic thesis posed by the telegraph and 3-D movies resulted in a refined thesis which tells more about the evolution of media than did the original thesis.

Of course, as Popper himself readily recognizes, a theory must first establish itself with affirming evidence before it can profitably seek out contradictions. "I realized [that] . . . somebody had to defend a theory against criticism," Popper writes, "or it would succumb too easily, and before it had been able to make its contributions to the growth of science."[22] Thus, the present inquiry should in no fundamental way violate the falsificationist approach by attempting to amass as much corroborating evidence as possible, even as it tries to scale the more intriguing "high altitudes" of anthropotropic evolution.

22. Karl R. Popper, *Objective Knowledge* (London: Oxford University Press, 1972), p. 30.

As yet a further expression of the falsificationist perspective — and as a logical consequence of an historical survey of media development followed by a series of evolutionary principles — the present study will conclude by hypothesizing the future course of media evolution, in effect applying the principles of anthropotropic evolution to the present state of media so as to predict the future. It is of course the essence of science that it generate predictions which are testable or capable of being proven false; yet it must be admitted that the predictions to be presented here will be largely immune to at least any immediate testing. Although predictions concerning the course of human affairs, of which the evolution of media is certainly a part, are indeed subject to an acid test, it is the acid test only of time, which is notoriously slow-acting and often inconclusive. If, for example, anthropotropic principles had in 1876 predicted the videophone to supplant the telephone, such a prediction would have yet, for reasons discussed in Chapter 7 below, to be fully vindicated or refuted. Similarly, a description of the telephone and automobile as grandparents to a line of evolution which will culminate in the instantaneous transmission of complete human beings across space — in effect achieving, as will also be discussed more fully in Chapter 7 below, a remerging of communication and transportation into a single act — would of necessity require many years to conclusively falsify. Thus, predictions and scenarios of future media developments will be offered here more in the spirit of logical speculation, than of strictly testable, scientific projection.

LIMITATIONS: WHERE THE THEORY NEEDN'T GO

Having attempted to outline what this study hopes to demonstrate, and how it will go about demonstrating it, this introductory chapter can perhaps further help define the purpose and method of the study by briefly identifying several areas that the study will not explore:

- Although there seems no obvious reason to suspect that anthropotropic evolution occurs only in communication technologies — indeed, recent developments such as the "biotecture" or living-form movement in architecture[23] suggest anthropotropic parallels in other technologies — the present study will concern itself only with the evolution of technologies used to transmit information or communicate. (An exception to this limitation is transportation technology, which, as suggested above, appears to bear a special relationship to communication media.) Again, there seems no reason, in principle, that should bar the constructs developed here from being fruitfully applied to other technologies, or to technology in g e n e r a l ; although it is conceivable that communications technologies are in the vanguard, and hence most observable section, of a general technological evolution towards production of human environments. In this regard, William Kuhns has noted that new technologies are usually first applied to communication rather than more "practical" tasks such as food-gathering and shelter, with the printing press predating the use of interchangeable, mechanized parts in heavy industry by at least 300 years, and the telegraph being the first major commercial application of electricity.[24] The primacy of communication technology would also be consistent with the view of George Herbert Mead, Martin Buber, and others who see the ability to communicate as the defining trait of humanity. [25]

23. Roy Mason, "Biological Architecture," *The Futurist* 11 (June 1977), 140-147.

24. *The Post-Industrial Prophets* (New York: Harper Colophon, 1971), pp. 153-154. Kuhns credits Harold Innis with making this observation; for more on Innis, see the end of Chapter 3 below. For more on the special status of communication technologies, see Chapter 4 below; but see also Chapter 4, n. 15 below for a suggestion of an "anthropotropic"-like model for the evolution of tools.

25. See George Herbert Mead, "Thought, Communication, and the Significant Symbol," and Martin Buber, "Between Man and Man: The Realms," in The Human Dialogue, eds. Floyd W. Matson and Ashley Montagu (New York: Free Press, 1967), pp. 397-403 and 113-117, respectively, for brief introductions to these thinkers.

- The perspective of the study will he that communication technologies evolve in response to human patterns of communication, with other human factors such as economics, politics, fashion, and the like being accounted for at most as either stimulating or inhibiting an inevitable course of evolution, or in some cases being the products themselves of underlying human communication preferences. The fact that the technological replication of motion preceded the replication of color, for example, may have bean due to the greater consumer demand or market for motion photography; but that greater market demand may itself reflect the greater importance of motion in the pre-technological communication environment (colorblind organisms have a lot better chance of survival than those that can't see motion). And in any case, the overriding point from the anthropotropic perspective is that both motion and color were eventually replicated, notwithstanding any differences in the economic climates, political receptivity, etc., at the time of their inceptions. Thus, the present study should be on safe ground in describing the evolution of media independent of economic and similar social factors, or with such factors, in effect, held constant.

- The thesis that media are evolving towards greater approximation of human communication systems argues only that media are functioning, or presenting information, in ways and forms similar to the way information is processed hy human systems; this suggests that media are becoming more human by performing more like human systems, and not necessarily by becoming physically more like human beings, or attaining the literal structure of human communication organs. In biology, this convergence of function without convergence of physical structure is found in such organisms as bats and birds, which achieve a comparable flight through differing physiologies;in media, the camera that captures color, for example, need not be any more physically like the living tissue optic system than the camera that captures only black-and-white — the resemblance between color camera and human eye arising here in the similar way that both systems process information or look at the world, that is, in the fact that both systems are sensitive to the spectrum of lightwaves that gives rise to the human experience of color. And since both systems process information similarly, it follows that the product or content of the color camera would resemble what we see (as far as color is concerned) in the real, pre-technological world. Thus, the anthropotropic trend may he described as a convergence of process or performance, resulting in a likeness between the products of media and human-perceived environments, without necessarily entailing a physical resemblance between the processors or performers themselves; the new contact lens described earlier in this chapter "breathes" like a living cornea, but is fashioned from inorganic plastic.

- In suggesting that media evolve towards greater replication of pre-technological reality, the anthropotropic thesis necessarily asserts that deliberate use of technologies to create "non-real" environments, as in art, seems in no way to divert the evolution of technologies towards greater reality replication. This is not to suggest that such non-real activities are unimportant to the human situation, but only that intentional distortion of reality appears not to be a primary goal cf technological evolution (except insofar as extension beyond biological boundaries may in itself be considered a "distortion" of pre-technological reality), and thus need not occupy a central portion of this study. Art, defined by numerous theorists such as Susanne Langer, E. H. Gombrich, and Jean-Paul Sartre as entailing a transcendence or "otherness" from reality, [26] is an especially interesting and significant example of such a non-real activity that plays an important role in human life, apparently without working against the evolution of media to more realistic forms. Possible explanations for this "co-existence" of non-real art with technological evolution towards the real are perhaps to be found in the observations of Mar shall McLuhan, who has noted that out-moded technologies often become admired as art, [27] or the work of film theorist Rudolf Arnheim, who suggests that earlier media such as silent movies that extend only single, "isolated" senses often make the best portrayers of art.[28] These views suggest that art does its hest with least realistic technologies, and that art, therefore, fails to deflect the evolution of media to more realistic forms because art, in effect, benefits from the cast-offs of such evolution. At the same time, however, philosophers have long recognized that for non-real art to be effective, it must maintain a firm grounding in reality — that Coleridge's "willing suspension of disbelief" must be coaxed by illusions that are believable, i.e., in some way resemble reality.[29] This view suggests a much more active, constructive relationship between art and anthropotropic technology.[30]

> But in either case, the continued existence, and importance, of art appears to pose no problem for the present thesis.

26. See, for example, Susanne Langer, *Feeling and Form* (New York: Charles Scribner's, 1953), E. H. Gombrich, *Art and Illusion*, rev. 2nd ed. (Princeton, N.J.: Princeton Univ. Press, 1972); and Jean-Paul Sartre, "The Work of Art," in *Aesthetics*, ed. Harold Osborne (London: Oxford Univ. Press, 1972), pp. 32-38.

27. Understanding Media: The Extensions of Man, 2nd ed. (New York: Mentor, 1964), p. ix. Examples would be horseback riding as a "sport" after the advent of automobiles, painting becoming more impressionistic once photography began doing a better job at portraits, and McLuhan's suggestion on p. ix that "the machine turned Nature into an art form." The process that seems to be at work here is that when a new technology replaces an old technology and therein "liberates" it from practical function, the old technology is freed to be appreciated not for what it does, but for what it "is," or as a "work of art."

28. *Film as Art* (Berkeley and Los Angeles: Univ. of Cal. Press, 1957), pp. 229-230.

29. Such a view was expressed in 1833, for example, by John Abercrombie, who wrote in *Inquiries Concerning the Intellectual Powers, and the Investigation of Truth* (original ed., n.p., 1833; reprint ed., Boston: Otis, Broaders, 1843), p. 129, that "a painter . . . depicts a landscape combining the beauties of various real landscapes . . . The compound in these cases is entirely fictitious or arbitrary; but it is expected that the individual elements shall be such as actually occur in nature, and that the combination shall not differ remarkably from what might really happen." (See Ch. 4, n. 49 below for a fuller presentation and discussion of Abercrombie's analysis of art.) The Coleridge quote comes from his 1817 *Biographia Literaria*, later edited by J. Shawcross (London: Oxford Univ. Press, 1907), vol. 2, p. 6.

30. Indeed, the view that effective art must not only transcend, but in some way continue to echo or reflect reality, suggests that art not

only directly benefits from anthropotropic media evolution, but in effect parallels media evolution and is a metaphor for it: for just as art attempts to go beyond yet retain a connection to reality, so advanced technology (as viewed in the present thesis) attempts to transcend the limitations of real time and space while retaining other elements of reality. The difference is that art's transformation of reality is conceptual while technology's is physical; and that art tends to emphasize its departure from reality more than its connection to reality, while technology seems increasingly to focus on the retrieval of reality, with the extension across time and space often held constant (as in the change from black-and-white to color TV). Since the present study is concerned with evolution of physical technologies and their distinctive relationship to reality, further theorizing about the process of art must be postponed for another time. Suffice it to point out for now, however, that in as much as both art and technology at once transcend and reflect reality, they exhibit in common the fundamental human quality (to even further paraphrase Carl Sagan) of being agents of nature, that go beyond nature, so as to examine nature. See also Ch. 4, n. 49 below.

- **Finally, it should be emphasized that the purpose of this study is to demonstrate and explore the evolution of media towards replication of human environments, and not necessarily to suggest whether such a pattern is desirable or dangerous to humanity. Thus, while it may seem unlikely, it is nevertheless possible that in recapturing elements of the human, pre-technological environment, media may be retrieving and amplifying those nether aspects of humanity which, as in Freud's "thanatos" and Jung's "shadow side" of human existence, engender despair and destruction. That the supplantation of artificial environments with those more consonant to living systems should fulfill some sort of primal "death-wish" indeed seems contradictory and unlikely — yet this, along with** other possible ill-consequences of anthropotropic evolution, will be briefly considered in the study's "Epilog" (Chapter 9), after the arguments for the existence of an anthropotropic pattern of media evolution have been presented.

In sum, then, this work will attempt to demonstrate, first through the perceptions of previous theorists, and then from observation of the past, present, and possible future of media themselves, that media are evolving towards an increasing replication of pre-technological or human communication environments. Such a pattern of "human replay" in media will be suggested: (1) primarily in communication technologies, and not necessarily in other technologies or technology in general; (2) notwithstanding the intervening influences of economics and other social factors; (3) as a resemblance of function and a resemblance of product, and not necessarily of physical composition or structure; (4) notwithstanding the usage of technologies for artistic and other "non-real" activities; and (5) independently of whether this convergence may work for the human detriment or good.

In turning to a consideration of previous thinking on the problem of media and related areas, it is well to bear in mind that no fully-developed, logically testable theories of media evolution are to be found here. The absence perhaps reflects the lack of scholarly interest in communication itself until little over thirty years ago, and the fact that communication technologies are in a still emergent, though certainly visible, state. What will be found are relevant fragments and analogies in other theoretical frameworks, and essentially poetic visions that attempt to make some initial sense out of the enormity, complexity, and sheer newness of communications technology. But from these shards and sinews of thought, the beginnings of an encompassing, unifying, and, one hopes, factually responsive theory of media evolution should herein take form.

PART I

ROOTS AND THREADS OF ANTHROPOTROPIC THEORY

CHAPTER 2

TECHNOLOGY AS EXTENSION

As suggested in the previous chapter, perhaps the most striking characteristic of communication technology to date — and hence its characteristic most likely to engage the attention of serious observers — is its extension of human processes, albeit via distortion, across space and time. Thus, while media observers may disagree about the impact and value of extension, nearly all have been united in the recognition that what technology does is allow us to "visit" times and places that were formerly unobtainable through unaided biology. The visit is possible because technology provides psychologically acceptable substitutes or extensions for human sensory systems.

SIGMUND FREUD: THE PROSTHETIC GOD

But the potent concept of extension did not originate with communication theory, and indeed has roots that reach well back into the 19th century.

Karl Marx's realization that such seemingly diverse entities as money, commodities, land, and machinery were all "transformations" or extensions of a fundamental human productive force, or "labor," precedes by nearly a century the explanation of communication technologies as extensions of human systems, and can thus be seen as the spiritual forebear of this type of "extensional" analysis. [1] But Marx never looked into the relationship of communication technologies and their individual human users, perhaps be cause at the time of his theorizing, communication technologies were still in their infancy, or perhaps because his concern was with the broad currents of society and history; and indeed, Marx apparently never regarded "transformation" as anything other than an equation involving vast, impersonal forces. It remained for psychologist Sigmund Freud to personalize the concept of extension, and apply it specifically to communications media.

1. The fullest expression of Marx's transformationalism of course

appears in *Capital*, first published in 1867. See Karl Marx, *Capital*, ed. Frederick Engels, 4th ed., edited, revised, and amplified by Ernest Unterman (New York: Modern, 1906), pp. 7-10 for a brief discussion of the publication dates and circumstances of various volumes of *Capital*. See also Robert L. Heilbroner's "Inescapable Marx," *New York Review of Books*, 29 June 1978, pp. 33-37 for a thorough summary and criticism of recent scholarship on Marx.

As a psychologist concerned with the individual, Freud was bound to think about technology in light of its relationship to human sensory and other biological systems. Moreover, as the developer of a theory that emphasized the passage of material between the unconscious and conscious, Freud was likely to rely heavily upon some mechanism of transformation or extension. (Indeed, had Marx not first devised it, Freud would surely have originated a postulate of transformation, for his psychoanalytic thesis required it.) Thus, in a paragraph in *Civilization and Its Discontents*, first published about 1930, Freud presents perhaps the first specific appraisal of communications technologies as extensions of human physiological and psychological processes:

> With every tool man is perfecting his own organs, whether motor or sensory, or is removing the limits to their functioning. Motor power places gigantic forces at his disposal, which, like his muscles, he can employ in any direction; thanks to ships and aircraft neither water nor air can hinder his movements; by means of spectacles he corrects defects in the lens of his own eye; by means of the telescope he sees into the far distance; and by means of the microscope he overcomes the limits of visibility set by the structure of his retina. In the photographic camera he has created an

> instrument which retains the fleeting visual impressions, just as a gramophone disc retains the equally auditory ones; both are at bottom materializations of the power he possesses of recollection, his memory. With the help of the telephone he can hear at distances which would he respected as unattainable even in a fairy tale. Writing was in its origin the voice of an absent person; and the dwelling-house was a substitute for the mother's womb, the first lodging, for which in all likelihood man still longs, and in which he was safe and felt at ease. [2]

2. New ed., translated and edited by James Strachey (original ed., 1930; new ed., New York: Norton, 1961), pp. 37-38. According to Strachey's Introduction, p. 5, *Civilization* was actually published in 1929, although the 1930 date is listed on the title page (Strachey cites Ernest Jones as the source of this information).

In this compact essay on technology, Freud makes at least four distinct and important observations: (a)industrial technologies extend human muscles, and transportation technologies extend human bodies across space, in ways very analogous to the operation of communication technologies; (b)communication technologies extend human senses beyond biological boundaries of time and space perception; (c)communication technologies extend not only externally-oriented physio-sensory systems such as sight, but also extend and reflect human internal psychological states such as memory and imagination; (d) the impetus for progressing technology comes from an inherent human desire to recapture a lost condition (and not, as Ellul might suggest, from a mesmerizing characteristic of technology that makes us do "its" own bidding).

The last two points are especially pertinent to an anthropotropic formulation. By relating the use of technology to primal human desires, Freud provides a motivation for the anthropotropic development. Why do we build technologies that increasingly recapture human communication environments? Why — because we love ourselves. It is not by choice that we currently live in Ellul's "lunar world of stone," only by necessity of our still primitive technologies. But just as life copies itself on a genetic level, human-produced technology seems to increasingly copy human communication on a cultural level.

Yet we know that technology not only copies but extends, and this extension — the instant traversing of time and space — is certainly not an obvious feature of the human, pre-technological environment, or "the mother's womb." What, then, is the motivation for extension? Freud's third point suggests an answer: in our mind's ever-potent eye and ear, we have all talked to people long dead or yet unborn, instantaneously traveled to far-off places on this globe and perhaps other planets, witnessed profusions of events perhaps "unattainable even in a fairy tale."

Thus, the motivation for technological extension is imagination — information traveling at the speed of light mimicking the speed of our thought, attempting through electronics and photo-chemistry to equal the awesome reach of our memory and anticipation, seeking to actualize our private capacity to communicate with anyone, anywhere, at anytime. Freud explains this as a fulfillment of a God-wish:

> Long ago he [humanity] formed an ideal conception of omnipotence and omniscience which he embodied in his gods. . . . Today he has come very close to the attainment of this ideal. . . . Man has, as it were, become a kind of prosthetic God. When he puts on all his auxiliary organs he is truly magnificent . . . . [3]

3. Ibid., pp. 38-39.

Of course, claiming the vast expanse of imagination as a determinant for media evolution runs the risk of opening a huge tautological gap in anthropotropic theory: if media evolve in "attainment" of imaginary ideals, then any and all manner of media may be explained and predicted as being merely imaginable — thus depriving the theory of any discriminating explanatory and predictive significance. The problem can be easily avoided, however, by recognizing that human imagination is not the totally unstructured source of possibilities we often take it to be; rather, "in the exercise of imagination," as 19th century philosopher John Abercrombie put it, "we take the component elements of real scenes, events, and characters, and combine them anew by a process of mind itself so as to form compounds which have no existence in nature."[4] In other words, the colors, sounds, textures, etc. that appear in our imagination are not "figments" of our imagination at all, but reflections, in one way or another, of the real world. The rearrangement and extension of these real-life constituents in our imagination, then, serves as a tangible, definable determinant for the extension of the real world through technology.

4. *Intellectual Powers*, p. 129. See also Ch. 4, n. 49 and Ch. 6 , n. 14 below.

Moreover, Freud's identification of imagination as the inspiration for technological extension reinforces one of the underlying perspectives of the present work, by suggesting that even the most primitive technologies — i.e., those that extend without replicating — attempt to fulfill a mandate that is fundamentally human, not machine-imposed. It becomes the task of anthropotropic theory, then, to trace how these imperfect extensions that gratify the imagination evolve to more perfect extensions that perhaps gratify the imagination even more.

Freud himself was the first to recognize that the imperfections of technology may be transitory:

> ... those [auxiliary] organs have not grown on to him and they still give him much trouble at times. Nevertheless, he is entitled to console himself with the thought that this development will not come to an end precisely with the year 1930 A.D. Future ages will bring with them new and probably unimaginably [or perhaps not so unimaginable] great advances in this field of civilization and will increase man's likeness to God still more. [5]

5. *Civilization*, p. 39.

Freud thus in several significant ways becomes the first anthropotropic theorist. That Freud's insight into the human/technological relationship has gone apparently unnoticed by subsequent students of media and technology is one of the more interesting oversights of media scholarship, perhaps the result of Freud's immense contribution and reputation in his own chosen discipline of psychology, tending to divert attention from his other accomplishments. It is certainly true that in the psychoanalytic perspective, technology serves as only one of many ways of wish fulfillment, and hence plays a commensurately peripheral role. Yet the import of even Freud's few words about technology shows that the throwaways of genius are often worth more than the labored treatises of ordinary men.

R. BUCKMINSTER FULLER: EYES EQUAL EYEGLASSES

Technology has heen more a passion than an aside for architect Buckminster Fuller — yet this hasn't prevented him from making important contributions to extensional theory, beginning with his *Nine Chains to the Moon* published eight years after Freud's *Civilization*, in 1938. Fuller's conception of technology starts here with his conception of the human body, which he sees as a grand machine:

> Man?
>
> A self-balancing, 28-jointed adapter-base biped; an electro-chemical reduction-plant, integral with segregated stowages of special energy extracts in storage batteries, for subsequent actuation of thousands of hydraulic and pneumatic pumps, with motors attached; 62,000 miles of capillaries; millions of warning signal, railroad and conveyor systems; crushers and cranes (of which the arms are magnificent 23-jointed affairs with self-surfacing and lubricating systems, and universally distributed telephone system needing no service for 70 years if well

managed); the whole, extraordinarily complex mechanism guided with exquisite precision from a turret in which are located telescopic and microscopic selfregistering and recording range finders, a spectroscope, et cetera, the turret control being closely allied with an air conditioning intake-and-exhaust, and a main fuel intake. [6]

6. *Nine Chains to the Moon* (Carbondale, IL: Southern Illinois Univ. Press, 1938), p. 18.

To which civilization has added "secondary mechanical compositions of its own devising, operable either by a direct mechanical hook-up with the device, or by indirect control through wired or wire-less electrical impulses"[7] — i.e., technology. All of which — the primary (bodily) as well as secondary (technological) components — would be an "imbecile contraption" without the guidance of an ineffable "phantom captain" or self,[8] the human mentality or soul that launches and pilots the mechanized ship in life, and abandons it in death.

7. Ibid, p. 19

8. Ibid.

Thus, our eyes, ears, and nervous system are themselves extensions in Fuller's model — extensions of the self — and as such eminently interchangeable, adaptable, and cooperative ("synergistic" in Fuller's later parlance) with the more detachable extensions of technology. What difference is there between an eye and an eyeglass, save that the eye is an extension of the mind more difficult to replace at the moment than the eyeglass? Thus the phantom captain readily uses "such service parts as crude gold inlays in 'his' raw fuel crushers, additional lenses or color-filters for 'his' range-finders, or an enema bag douching nozzle temporarily passed into 'his' clogged canal. The inlay or douche bag is, temporarily at least, as factually connected to self as toenail, tooth, hair or eyeball."[9]

9. Ibid, pp. 21-22.

In this analysis, Fuller has in effect pursued the concept of technological extension to its extreme, by making technology a functional equivalent of human biological systems. This equation of flesh-and-blood with steel- and-wire supports the anthropotropic thesis in at least two ways: (a) If technology and the human body are indeed interchangeable extensions of the mind or self, then it stands to reason that the mind would direct technology to create environments compatible with both types of extensions, which in effect means an environment compatible with the more vulnerable flesh-and-blood components of the human body. For the mind to direct technology to do otherwise would be for the mind to set the two types of extensions at war with each other, a development which would probably result in the destruction of both. (Critics of technology of course see just such a conflict between human and technological equipment — but such conflict is illogical and unmotivated in Fuller's framework.) (b) As an ultimate development, the mind might be expected to create technologies so interchangeable with flesh-and-blood that the technologies were themselves flesh-and-blood. But the literal duplication of human sensory systems rather than the processes by which sensory systems operate is an occurrence which, for reasons explained m the first chapter, the anthropotropic theory need not predict. Fuller's main relevance to anthropotropic theory, then, lies in the vivid parallel that he draws between biological and technological means of communication (a parallel subsequently explored in greater depth by cybernetic theory, discussed in Chapter 4 below).

As might be expected, Fuller's views have been roundly criticized by most opponents of technology, who see in his metaphor for human life the epitome of dehumanized, mechanized existence. Thus Lewis Mumford, who will be heard from in further detail in the next chapter, comments in his typically dyspeptic way that had Fuller not originated the "28-jointed" description of man, " I might be accused of having wantonly invented [it] for the purpose of exposing the crudity and absurdity of the original doctrine."[10] What Mumford and others have apparently missed, however, is Fuller's insistence on the "phantom captain" or self as the essential force that makes life meaningful, indeed possible ("without which human mechanisms are imbecile contraptions"). The worst, perhaps, that Fuller can be justly accused of is an "old-fashioned" mind/body dichotomy — for which, at the risk of opening a tangential can of worms, it must be said there is still much evidence and certainly no convincing disproof.[11]

10. *The Myth of the Machine, vol 2. The Pentagon of Power* (New York: Harcourt Brace Jovanvitch, 1970), p. 56.

11. See W. H. Thorpe, "Arthur Koestler and Biological Thought," in *Astride the Two Cultures*, ed. Harold Harris (New York: Random House, 1976), pp. 50-68, for a brief summary of the arguments for the existence of mind. Karl R. Popper and John C. Eccles offer a fuller treatment of mind/body dualism in *The Self and Its Brain* (New York: Springer International, 1977).

EDWARD HALL: SPIDERS' WEBS AND TV NETWORKS — A BIOLOGICAL PURPOSE FOR EXTENSION

The next step in explicating the process of technological extension comes from cultural anthropologist Edward Hall, who draws a parallel between the spider's web, the beaver's dam, and human technologies — all specialized extensions that help the organisms survive in their environments. Writing some twenty years after Fuller's first publication, Hall in *The Silent Language* observes that:

> Occasionally organisms have developed specialized extensions of their bodies to take the place of what the body itself might do and thereby free the body for other things. Among these ingenious natural developments are the web of the spider, cocoons, nests of birds and fish. When man appeared with his specialized body, such extension activities came into their

own as a means of exploiting the environment.

Today man has developed extensions for practically everything he used to do with his body. The evolution of weapons begins with the teeth and the fist and ends with the atom bomb. Clothes and houses are extensions of man's biological temperature-control mechanisms. Furniture takes the place of squatting and sitting on the ground. Power tools, glasses, TV, telephones, and books which carry the voice across both time and space are examples of material extensions. Money is a way of extending and storing labor. Our transportation networks now do what we used to do with our feet and backs. In fact, all man-made material things can be treated as extensions of what man once did with his body or some specialized part of his body.[12]

12. (New York: Fawcett, 1959), p. 60.

In his comprehensive summary of technological extension that encompasses even the economic transformationalism of Marx, Hall discovers that technological extensions are biologically adaptive. By noticing that human technology and the spittle of insects share a common genetic purpose — to better cope with the environment — Hall gives technological extension a biological raison d'etre comparable to the psychological motivation for technology suggested by Freud.

The identification of technology as an adaptive mechanism strongly implies an anthropotropic development of technology. There certainly seems no adaptive dividend in extensions which distort non-threatening elements of the natural environment, as in black-and-white TV and blind telephone. Indeed, it would seem that the greatest adaptive advantages would lie in those extensions which transcended certain unwanted limitations of the environment, while conserving the remainder of the environment in which humans feel at home. Which is precisely what would be expected on the basis of anthropotropic theory.

On the other hand, Hall's reference to the atom bomb raises a discordant note which can hardly be ignored: it is palpably obvious that the bomb is less biologically adaptive — i.e., more of a danger to the survival of the species — than earlier extensions like the rifle, or pre-technological weapons like fists and teeth. The problem no doubt stems from technology's capacity to extend — and retrieve — elements of the human pre-technological environment which are no longer adaptive, or which were useful in their original pre-technological state, but counter-productive in extensional form. Thus, many anthropologists have pointed out that human aggression was a more appropriate response to bygone environments[13] and the fist certainly seems more adaptively flexible than the bomb. Such facts of life and extensions don't necessarily contradict the anthropotropic thesis. But they raise serious questions about the ultimate benefits of technological evolution, which will be further considered in the "Epilog" of this study.

13. See, for example, Konrad Lorenz, *On Aggression* (New York: Bantam, 1963).

MARSHALL MCLUHAN (I): HUMANS AS TECHNOLOGICAL EFFECTS

Consideration of the impact of technological extension introduces Marshall McLuhan, whose name justifiably has become synonymous with the very study of communications media. Alone among the theorists thus far discussed — and, for that matter, to be discussed — McLuhan has devoted the full force of his attention and ingenuity to an examination of communication technologies, producing a body of work rife with insight and possibility that has legitimized media as a subject for serious inquiry in philosophy and the humanities. Although McLuhan's writing has frequently been derided as speculative and recklessly stylistic,[14] critics usually miss the point that McLuhan's speculations are informed with an almost encyclopedic knowledge of communication systems ranging from Greek poleis to medieval monasteries. Moreover, it is altogether possible that McLuhan's aphoristic net is the only sure way to catch the big fish in a maiden voyage across a murky media sea. In any case, while the present inquiry may disagree with some of

McLuhan's perspectives, it draws upon McLuhan's thinking as a continuing source of new ways of looking at new and old things. The following discussion, of necessity, touches upon only the highlights of McLuhan's work that appear to have the most relevance to an anthropotropic theory of media.

14. See, for example, the untitled essays hy Dwight Macdonald and Ben Lieberman in *McLuhan: Hot and Cool*, ed. Gerald Emanuel Stearn (New York: Dial, 1967), pp. 203-211, and 222-225, respectively.

Much like the present inquiry, McLuhan recognizes the ability of technology to extend human systems, and views this as but a first step in achieving a more encompassing understanding of communication technologies. But unlike the anthropotropic formulation, which sees the next step as investigating how these extensions come to conform to human environments (i.e., Stage C), McLuhan investigates the reverse, and wonders how the human environments come to he remade in the image of the incomplete extensions. The basis of such an investigation resides in McLuhan's belief that extensions have the peculiar property of transferring power from origin to extension, putting the car or road in the driver's seat, as it were, and leaving the hapless driver at the mercy of his former "servants." McLuhan explains this power-switch with the example of the Greek youth Narcissus, who "mistook his own reflection in the water for another person. This extension of himself by mirror numbed his perceptions until he became the servo-mechanism of his own extended or repeated image." [15] In other words, our failure to re cognize extensions as parts of ourselves invests extensions with autonomous power, to perhaps control us.

15. *Understanding Media*, p. 51. More recently, Hall too has noticed a tendency of humans to reverse places with their extensions. He writes in *Beyond Culture* (New York: Anchor, 1976), p. 25, that, "extension transference — ET — is the term I have given to this common intellectual

maneuver in which the extension is confused with or takes the place of the process extended . . . . Worshipping idols, common to all cultures, represents one of the earliest examples of the ET factor. In the Bible, we see this when men are directed to give up the worship of 'graven images.'"

Such extensional or media determinism when taken to its extremes begins to resemble Ellul's view that our technologies are enslaving us.[16] Indeed, McLuhan can on occasion be so extreme in the media determinist perspective that he borrows Samuel Butler's famous observation about the chicken-being-the-egg's-way-of-producing-more-eggs to characterize the human role in technological evolution, to wit, "man becomes, as it were, the sex organs of the machine world . . . enabling it to fecundate and to evolve to new forms." [17] Although such an assertion is no doubt intended to call attention to the important degree to which humans are products of their own technological creations, the implications of McLuhan's assertion go much further, depicting humans as the passive, mindless partner in the human/technological relationship. This view is obviously at some variance with the anthropotropic perspective which, while recognizing the two-way relationship between human creators and human creations, suggests that human creators continue to direct the course of the relationship. This seems to follow from, if nothing else, the very nature of the terminology used to describe the human/ technological relationship: for while a product may indeed influence its producer, it is the producer that is the primary source of production, that is, it is the producer, not the product, that produces.

16. As suggested on p. 3 above, however, Ellul's pessimism stems from the quite different notion that it is the human impulse for organization which, extended or not, leads to our predicament with technology. The view of humans as servants of machines of course predates both Ellul and McLuhan, and goes back at least as far as Marx. For more on Ellul's

perspective, see *The Technological Society*, pp. xxv-xxvi, 24-26; for Marx's view of workers as servants of machines, see *Capital*, p. 461.

17. Understanding Media, p. 56. Butler's observation about the primacy of the egg comes from his Life and Habit (original ed.,n.p., 1878; reprint ed., New York; Dutton, 1910), p. 134; "It has, I believe, been often remarked, that a hen is only an egg's way of making another egg." Butler was also a very early and remarkably prescient thinker about technologies, writing in 1863 and 1865 two letters which anticipated nearly every observation and argument that would be made about technology in the twentieth century. In the first letter, which now reads like a caricature of positions seriously taken by Jacques Ellul and other critics of technology, Butler asks "what sort of creature man's next successor in the supremacy on the earth is likely to be," and suggests that it will be "machines"; "an entirely new kingdom ... of which we as yet have only seen what will one day be considered the antediluvian prototypes of the race .... Day by day, however, the machines are gaining ground upon us; day by day we are becoming more subservient to them; more men are daily bound down as slaves to them, more men are daily devoting the energies of their whole lives to the development of mechanical life." Butler/ however, presents a more "anthropotropic" appraisal of technology in his second letter, suggesting that what distinguishes humans from the rest of the animal world are our "extra-corpotaneous limbs ... every fresh invention is to be considered as an additional member of the resources of the human body; ... in the deliberate invention of such unity of limbs as is exemplified by the railway train — that seven-leagued foot which five hundred [men] may own at once — he [man] stands quite alone.... If it is wet he is furnished with an organ which is called an umbrella .... His memory goes in a pocket [portable] book." Moreover, Butler recognizes that these extended "limbs" evolve in a Darwinian-like process of human selection: "Our ancestors added these things to their previously existing members; the new limbs were preserved by natural selection, and incorporated into human society; they descended with modifications, and hence proceeds the difference between our ancestors and ourselves." Butler's observations thus prefigure not only the views of technological extension provided by Freud, Fuller, Hall, and McLuhan — every theorist

in this chapter — but the views of theorists such as Peter Medawar and Karl Popper, to be discussed subsequently, which develop the notion that human products evolve like living organs. Butler himself, unfortunately, did little more to develop these ideas, although he did work the substance of the two letters into "The Book of the Machines'" section of his novel *Erewhon*, first published in 1872 (reprint ed., London: Dent, 1965). The two letters, titled "Darwin Among the Machines" (1863) and "Lucubratio Ebria" (1865) were first published in a New Zealand newspaper (*The Press*), and have since been reprinted in *The Note-Books of Samuel Butler*, ed. Henry Festing Jones (New York:AMS, 1968), pp. 35-46.

Moreover, the actual evidence of technological evolution, to be reviewed at length in Part II of this study, appears to support the anthropotropic perspective in this matter. In the first place, if new media were indeed created by humans at the psychological behest, as it were, of older media, then one might expect these new media to more or less continue in the same pattern as the older media. But as McLuhan himself was one of the first to recognize, new electronic technologies are in many ways diametrically the opposite of older mechanical print and scribe media (see next section on McLuhan, immediately following the present section, for discussion of McLuhan's views on print vs. electronic media). Are the electronic technologies, then, the "result" of older print media? Yes, hut only in the reactive sense that the human spirit, unsatisfied with the environment of print (note: actively unsatisfied, not mesmerized) strove to invent the quite different environment of electronics and photo-chemistry. This is human determinism, not media determinism.

But making technology the cause, and humanity the effect, seems to violate the evidence in perhaps a more profound way, for it denies the undeniable fact that humanly intelligent life must have existed before humanly produced technologies. For although language and tools no doubt were

hugely important in the refinement, and even development, of humanity, there can be even less doubt that some living intelligence had to first exist to invent and learn how to use the language and tools. This is one case where the chicken, of necessity, came before the technological egg. (The source of humanity would then perhaps lie in the non-technological realm of genetic mutation, divine intervention, etc.)

But if the media determinist perspective is unfit as a model of media evolution (and indeed, McLuhan would perhaps suggest that his perspective, at least, was never intended as such), [18] it is nonetheless useful for calling attention to the effects that media may have upon us as they evolve. To point out as McLuhan does in *The Gutenberg Galaxy*, for example, that "the extension of one or another of our senses by mechanical means, such as phonetic script, can act as a sort of kaleidoscope twist of the entire sensorium,"[19] is in no way to contradict the anthropotropic thesis that as technologies evolve, they become more complete — or less disruptive — extenders of the senses. Neither is there any necessary incompatibility between the present perspective and McLuhan's general observation that print, in extending a single, abstract way of seeing on an unprecedented,worldwide basis, unhinged human "sensory ratios" to encourage everything from the Scientific Revolution to schizophrenia. Indeed, it seems thoroughly reasonable to suggest that primitive Stage B technologies, which are capable of extending one sense only at the expense or the distortion of others, may profoundly alter their human users — though apparently not enough to preclude the human invention of subsequent technologies which are not as disruptive. (Note, incidentally, that primitive, distortive technologies do not end with print, but continue in a diminishing degree with electronic media such as the telephone, which extends the voice/ear at the expense of the face/eye.)

18. "I don't explain — I explore," McLuhan advises in *McLuhan: Hot and Cool*, ed. Stearn, p. xiii.

19. (New York; Mentor, 1962), p. 70.

Herein, then, is an avenue of reconciliation between media determinism and anthropotropic theory: the more primitive the technological extension (i.e.. Stage B) , the less real the created environment, and the more disruptive of human sensory ratios; and conversely, the more advanced the technological extension (i.e.. Stage C), the more real or complete the created environment, hence the less disruptive of human systems. Such a sliding scale plausibly equates balanced human sensory ratios with pre-technological reality (that is, communication in the pre-technological world of Stage A entails a non-extended biological balance of human senses), suggests that primitive extensions alter the sensory balance by extending one or two but not all the senses, and further suggests that subsequent technologies tend to restore the lost balance by extending a greater spectrum of senses. But though advanced technologies diminish sensory distortion, they as yet still fall short of total sensory extension, and hence the media determinist is entitled to ask: what does the shortfall in sensory extension of even advanced technologies do to the user? Holography, for example, may be explained by anthropotropic theory as a restoration of the third dimension previously missing in sensory extensions, and at the same time queried by media determinism as to what consequence the absence of its still missing extensions — such as tactility, smell — will have for its beholders. Moreover, the continued co-existence in our society of primitive as well as advanced technologies indicates additional areas for media deterministic exploration, which will be discussed more fully later in the study (see "Persistence of Abstraction" in Chapter 6).

MARSHALL MCLUHAN (II): ORAL, WRITTEN, AND ELECTRONIC — AN ANTHROPOTROPIC-LIKE SCHEMA

McLuhan actually comes close to suggesting the above distinction between primitive and advanced technologies, when he points out the differences between print and electronic media. Print and phonetic writing, as mentioned earlier, are described by McLuhan as isolating, segmented media, which magnify one sense — linear visuality — completely out of proportion to its position in the pre-technological human sensorium. Electronics, on the other hand, McLuhan considers a reintegrative, multi-dimensional force, helping literally to restore the color and diversity of sensation that have been bleached by print.

Thus, McLuhan titles the chapters of *The Gutenberg Galaxy* with such "glosses" as: "The interiorization of the technology of the phonetic alphabet translates man from the magical world of the ear to the neutral visual world," and "The increase in visual stress among the Greeks alienated them from the primitive art that the electronic age now reinvents after interiorizing the 'unified field' of electronic all-at-onceness," [20] and more recently explains that,

> Today, in the age of secularization of science, we have, by electronic means, recreated the conditions for the dominance of the right, or mystic, hemisphere. The environment of simultaneous information which now prevails electronically over the planet, once more favours the tribal and corporate and intuitive dominance of the right, or mystic hemisphere. [21]

20. Ibid., pp. 345-346.

21. Marshall McLuhan, review of *The Origin of Consciousness in the Breakdown of the Bicameral Mind*, by Julian Jaynes, Centre for Culture and Technology, University of Toronto, Toronto, Canada, 3 June 1977, p. 2.

McLuhan's three-part configuration of tribal, scribal-print, and electronic technologies — which, unlike the "egg-and-chicken" metaphor discussed above (p. 56), describes the development of technologies without reference to their parentage — in many.ways parallels the three-stage model proposed by the present inquiry. Thus, McLuhan's tribal world is non-technological, [22] as is the environment of Stage A, and the only extensions are the biological devices of imagination and voice; the invention of phonetic writing and then print allows communication beyond the biological boundaries of space and time — but the extension is one-dimensional, as are the primitive extensions of Stage B, in that users can "travel" beyond biology only by leaving what now becomes the "excess baggage" of their multi-sensorial world behind; the eventual development of electronic technologies reclaims the lost baggage of tribal communication, recreating an environment of "total" sensory involvement that aptly describes the face-to-face communication increasingly retrieved by Stage C technologies.

22. Except in so far as speech itself may be considered a "technology," see the end of Chapter 6, below.

Indeed, McLuhan's schema seems to coincide with the anthropotropic model in all but two minor respects: (a) The present inquiry sees in print (as opposed to pre-mass phonetic writing) the beginnings of the retrieval of pre-technological reality (see Chapter 5 below) , and at the same time observes that some very early electronic technologies, such as telegraphy, are in some ways more distorting of reality than print (see Chapter 5 below). In other words, the demarcation between Stages B and C of the anthropotropic model appears not to coincide precisely with the shift from print to electronics emphasized by McLuhan. (b)While McLuhan focuses on "simultaneity" of electronic environments as responsible for retrieving the multi-sensory tribal experience, the present study also emphasizes the ability of advancing technologies to actually replicate elements of the pre-technological (tribal) world, such as color, sight-and-sound, etc.

Upon closer examination, however, it turns out that electronic simultaneity may be retrieving the process, and technological color, sight-and-sound, etc. the content, of a same pre-technological environment. Moreover, this process and content seem integrally interrelated, since (i) technological production and conveyance of even two-dimensional images allows a Gestalt-like, simultaneous perception characteristic of pre-technological real world vision and absent in sequential writing; (ii) technological re-unification of sight-and-scund corresponds to the simultaneity of sight-and-sound in the real world; (iii) access of one perceiver to multiple sensory and informational sources through advanced technology (both intramedia — i.e., many possible programs at once on TV — or inter-media) corresponds to the real world, where a variety of possible or actual experiences may be available to a perceiver at any one time; and (iv) availability of one technological experience to many simultaneous perceivers, as in the nightly newscasts, corresponds to the pre-technological real world, where an event may be witnessed by all within biological proximity (as McLuhan has pointed out, privacy is an invention of literacy; see Chapter 5 below for further discussion). In short, simultaneity seems on all levels to be a prime organizing principle of pre-technological communication; McLuhan's emphasis of simultaneity as the key element in the electronic retrieval of the tribe thus in no way contradicts the thrust of the anthropotropic thesis.

In sum, then, McLuhan's three-part taxonomy of oral/written/and electronic technology — which, consistent with most of McLuhan's work, is more a series of suggestions than a formal taxonomy — is in most ways an important prefigurement of the three-stage model of anthropotropic evolution. Coming as it does from the different perspective of media determinism, such agreement lends convincing support to the anthropotropic theory.

As a final note in the consideration of McLuhan, for the present, it should he pointed out that McLuhan is very much aware, as is this inquiry, that the electronic retrieval of reality differs from pre-technological reality in at least one crucial respect: electronic retrieval confers upon the perceiver the potent capacity to instantly transcend space and time — a capacity present in the pre-technological world only in the non-material imagination. McLuhan is especially worried about the vanquishing of space, and writes that,

> At the speed of light, minus his physical body, man is discarnate ... This anarchic elevation of nuclear man enables individuals to be dispensed, as it were, from the moral law, a fact which was strikingly manifested in the radio age by Stalin and Hitler, and in the TV age by the universality of abortion.[23]

McLuhan's misgivings should perhaps be allayed by the fact that electronic technologies now not only extend across space but across time — that is, they permanentize as well as quicken — and thus provide an important balance to the loss of distance (see discussion of Harold Innis in Chapter 3 below). Or it might be observed, as was earlier, that information traveling at the technological speed of light only mimics information traveling at the human speed of thought. But such observations don't really obviate the enduring difference between the limited reality of biology and the unlimited reality of advanced technology, or the difference between imagination and imagination-instantly-materialized. And thus the enduring difference between anthropotropic theory and McLuhan's thinking may well lie in the first investigating how technologies produce environments that are inevitably more human, while the second investigates how these environments are inevitably more (or less) than human.

23. "Note on Discarnate Man," Centre for Culture and Technology, University of Toronto, Toronto, Canada, 10 May 1977.

Investigations of the first type, till now, have been far less common than investigations of the second; tout the second type may yet prove of more ultimate value to humanity.

Thus, McLuhan's work, which as indicated earlier has only been selectively discussed here, seems to offer the following points of support and conflict for an anthropotropic theory of media evolution: (a) Both frameworks recognize the extensional nature of technologies, and view this as but a first step towards a more in-depth investigation of technologies. (b) McLuhan is interested in the ways that technological extensions of human systems alter those systems. This "media determinist" perspective suggests, in the extreme, that humans are the "effects" of technologies, and that technologies evolve not in response to biological human systems, but in response to systems laid down by prior technologies. Such a view contradicts both the anthropotropic thesis and the logical absolute that humans (or intelligence) per force predate technology. In its less extreme expressions, however, the media determinist view that technologies alter human systems presents no necessary contradiction to anthropotropic theory; rather, the media determinist concern can be integrated into anthropotropic theory with the recognition that as technologies evolve, i.e., more increasingly retrieve reality, they grow less disruptive or distortive of biological sensory systems. (c) McLuhan1s historical division of communication technologies into tribal-oral, written, and electronic environments in many respects corresponds to the anthropotropic three-stage model of non-technological, early technological (extension at the expense of reality), and advanced technological (extension with reality retrieved) environments. As such, this is the specific thread of McLuhan's thinking that probably provides the strongest support for the anthropotropic thesis.

***

McLuhan's is certainly the most penetrating and elaborate analysis of extensional technology to date, and can thus be considered an apex of a train of thought that began over a century ago with the transformational theory of Marx.

To briefly recapitulate, that train has been fueled along

the way by theorists such as: Freud, who first identified technologies as extensions of human systems, explained how they gratified primordial human desires, and recognized their extension not only of exterior senses but of interior imagination; Fuller, who viewed both technology and the human body itself as extensions of the mind or self, thus demonstrating no necessary conflict between biology and technology; and Hall, who showed that technological extensions are biologically adaptive or useful for survival, helping organisms like spiders and humans better cope with their environments. All of which makes good grist for the anthropotropic mill.

As suggested earlier, however, it is noteworthy that none of these perspectives fully addresses the question of how technological extensions evolve, or in what direction, if any, such evolution is headed. In a sense, then, these "first generation" extensional theorists may be considered the Aristotles and Linnaeuses of technology — performing the important preliminary tasks of defining, observing, and identifying, and in McLuhan's case, of cataloging the consequences, of extensional usage.

As also suggested earlier, it is perhaps the still emergent state of much communication technology that has made a framework of media evolution so difficult to formulate. To find the few and fragmentary frameworks that have been suggested, however, it will be necessary to leave the psychologist, the architect, the cultural anthropologist, and even the media explorer, and consult instead the longer perspectives of the biologist and the historian.

CHAPTER 3

INTIMATIONS OF MEDIA EVOLUTION

"The past two centuries have seen an amazing amount of empirical evidence amassed to substantiate the doctrine of organic evolution," H. James Birx writes in *Teilhard de Chardin's Philosophy of Evolution*, adding that, "evolution has even been extended to include galactic, solar, geological, social, and psychological phenomena as well as biological species."[1] And what explanations have been put forth, till now, about the evolution of the biological/psychological/social phenomenon of technological extension?

1. (Springfield, IL.: Thomas, 1972), p. ix. See "Lamarck, Darwin, and Teilhard" later in the present chapter, and "Teilhard's Noosphere" in Chapter 4 below, for discussions of Teilhard's relevance to anthropotropic theory.

PETER MEDAWAR: AGENDA FOR A THEORY OF TECHNOLOGICAL EVOLUTION

"Everyone has observed with more or less wonderment," notes biologist Peter Medawar, "that the tools and instruments devised by human beings undergo an evolution themselves that is strangely analogous to ordinary organic evolution . . . . " Medawar then goes on to make his own astute observation that

> Aircraft began as birdlike objects but evolved into fishlike objects for much the same fluid-dynamic reasons as those which caused fish to evolve into fishlike objects. . . . The assimilation of technology to ordinary organic evolution has substance, because all instruments are functionally parts of ourselves. Some instruments like spectrophotometers, microscopes, and radio telescopes, are sensory accessories inasmuch as they enormously increase sensibility and the range and quality of

sensory input. . . .[2]

2. "What's Human About Man Is His Technology," *Smithsonian*, 4 (May 1973), p. 22.

Although Medawar discerns several "striking" differences between the evolution of biological organs and their technological extensions— which stem from the fact that biological organs are the immediate product of genes which relate to the environment in a naturally selective or Darwinian way, while technological extensions are constructed from cultural instructions that are transmitted in more of an additive or Lamarckian manner— he concludes that

> there are certain obvious parallels between exomatic [technological] and ordinary organic evolution in the Darwinian mode. Consider again, for example, the evolution of aircraft and of automobiles. A new design is exposed to pretty heavy selection pressures through consumer preferences, 'market forces' and the exigencies of function, by which I mean the aircraft must stay aloft and the cars must go where they are directed. A successful new design sweeps through the entire population of aircraft and automobiles and becomes a prevailing type much as jet aircraft have replaced aircraft driven by propellers.[3]

3. Ibid., p. 27.

Moreover, such changes, Medawar reminds us, "do not exemplify and are not the product of the workings of great impersonal historical or sociological forces." Rather, technological extension is a direct "human artifact; so also are the laws and regulations which govern its transformations."[4]

4. Ibid.

Medawar has thus provided a description of what a theory of technological evolution must do — a job description that in many ways anticipates the anthropotropic theory, in suggesting that: (a) the nature of technology is extensional, and the relationship between extension and human system is integral and interchangeable, along the lines described by theorists summarized in the previous chapter; (b) the mechanisms of technological evolution in many respects resemble the Darwinian model; (c) surviving technologies must perform the dual tasks of "functioning" and satisfying "consumer preferences," which in anthropotropic terminology roughly translates as extending across time and space (i.e., functioning), and retrieving pre-technological reality (i.e., satisfying consumer preferences); and (d) the on-going activation of the evolutionary process is not autonomously technological or Ellulian, but continuously human.

LAMARCK, DARWIN, AND TEILHARD: MODELS OF EVOLUTION

The point in Medawar"s program least discussed thus far in the present inquiry is the suitability of existing models of evolution for explanation of technological change. Although the investigation of anthropotropic evidence in ensuing chapters will no doubt provide a better basis to judge existing evolutionary models and construct a new one more tailored to technology, it may be useful at present to very briefly describe the Lamarckian and Darwinian frameworks mentioned by Medawar, and assess their applicability to anthropotropic evolution. In their simplest expressions, both models view evolution as an interrelationship among three entities: organism, environment, and transmitting agent (the third being the connection between organism and offspring: "genes"). Both models are substantially in agreement as to the relationship between organism and environment, namely, that organisms either adapt to the environment or die, and that evolution is a process of organisms attaining successively better correspondence or adaptation to their environments. The models disagree, however, on the relationship between environment and transmitting agent (and between organism and transmitting agent): In the Lamarckian model, the environment works directly on the transmitting agent— organisms respond to the environment, incorporate those responses into the transmitting agents, which produce offspring better adapted to the environment by reason of these incorporated responses. In contrast, the more complex Darwinian model[5] suggests that the transmitting agents operate more or less independently of the environment— "blindly" producing organisms with a variety of traits or responses, which are then "selected" by the environment for survival, on the basis of how well these "haphazardly" generated responses anticipate the environment. Although the Darwinian approach leaves some nag-ging loose ends — such as how an essentially trial-and- error, disjointed process can result in a vastly complicated biological world in intricate correspondence to its equally complicated environment— it has become the stand-ard model for organic evolution.[6]

5. Many modern biologists identify their evolutionary perspective as "neo-Darwinian" or "post-Darwinian," so as to distinguish their view from the original Darwinian model, which was postulated before the discovery of genetic mechanisms. However, since the present study will for obvious reasons use the Darwinian model as a metaphor rather than a literal and complete description of media evolution, the plain term "Darwinian" will be used here for the sake of simplicity. For more on evolutionary terminology, see Edward O. Wilson, *Sociobiology* (Cambridge, Mass.: Belknap, 1975), pp. 63-64.

6. See Thorpe, p. 61, and Arthur Koestler, *Janus: A Summing Up* (New York: Random House, 1978), pp. 165-192 for a discussion of some of the major inadequacies of Darwinian (and neo-Darwinian) theory. Koestler's main criticism of the Darwinian model, interestingly, is that it fails to account for the growing complexity of organisms— that is, it denies a pervading purpose or destination for organic evolution— which, as will be seen on at the of Chapter 3below, is the main inadequacy in the application of

the Darwinian model to media evolution. But see also Stephen Jay Gould, "Koestler's Solution," review of *Janus: A Summing Up* by Arthur Koestler, in *The New York Review of Books*, 20 April 1978, pp. 35-37, for a rebuttal of many of Koestler's criticisms.

In the realm of technological evolution, of course, technologies may themselves be considered the "organisms." And, as suggested in the Introduction, the environment to which technologies must respond is the environment of human preference. (Note that while Medawar is correct in observing that technologies must perform in the same outside physical environment as biological organisms — see his analogy above between aircraft and fish— technologies, unlike most biological organisms, survive only through the human appraisal of their performance in outside environments. Hence, the effective determining environment for technologies is human response.) What, then, is the "transmitting agent" for technologies? Since even the most sophisticated technologies, as yet, are incapable of reproducing themselves, it is obvious that the technological transmitting agent or "gene" is, once again, the human organism. (And in this limited sense, McLuhan's view that human beings are the "sex organs" of technologies would be applicable. But rather than implying an abject technological determinism, humans functioning as technological "genes" are both autonomous and creative, as will be explained immediately below.) The double human role as both environment and transmitting agent for technology derives from the human capacity to both use or consume technologies (the human user is the environment) and produce or create technologies (human production is the transmitting agent).

With such analogies in mind, it is easy to see how technological evolution can be both Lamarckian and Darwinian at times. When humans deliberately produce new technologies to redress inadequacies in the performance of prior technologies, and such new technologies are utilized as intended (i.e., to replace the prior technologies that performed inadequately in their environment), then these technologies may be thought of as evolving in a "Lamarckian" way: the human transmitting agents have responded directly to the requirements of the environment, and have produced technological "organisms" that attempt to satisfy these environmental pressures. But when new technologies are happened-upon "accidentally" — or when technologies deliberately developed in response to one environment wind up being utilized in another environment — then the evolution is certainly more Darwinian, for the human transmitting agents are producing technological organisms in independence of the environments the technologies will function in, and as such the human transmitting agents are behaving as autonomously as Darwinian organic genes.

Arthur Koestler has perceptively chronicled the central role that "discovery by misadventure" has played in the genesis of great ideas and inventions— the host of "sleepwalkers," from Pasteur to Faraday, from Edison to Roentgen, whose "discoveries were the happy outcome of a comedy of errors," the products of "that remarkable form of blindness which often prevents the original thinker from perceiving the meaning and significance of *his own* discovery [italics in original]."[7] Koestler cites such media as photography, x-rays, and sound recording as the progeny of such fortunate accidents; and to that list one can add numerous other communication technologies — such as Bell's inspiration for the telephone while in pursuit of a better hearing aid — which were similarly stumbled upon. Indeed, the very long and slippery path of technological evolution itself — riddled with false starts and failed gadgets, and premature births such as Leonardo's — suggests that direct Lamarckian progress may play a trivial role in the refinement of technologies, while the more important developments are best explained by the indirect but richer mechanism of Darwinian change.

7. *The Act of Creation* (New York: Macmillan, 1954), pp. 192, 214, 216 ff. Koestler first used the term "sleepwalkers" in his earlier book by that name, *The Sleepwalkers* (New York: Grosset & Dunlap, 1959).

There is, however, one serious respect in which the Darwinian model does not seem to adequately explain the evidence of technological evolution, at least in so far as that evidence is being assembled in the present work. Although the environment, in the Darwinian conception, is selective and hence providing of some direction to evolution, the exact course and eventual outcome of that direction are essentially unknowable in the Darwinian realm: the environment of any single organism is an unstable composite of all other organisms, in itself evolving so that its selective impact is all but impossible to specifically predict. In philosophic terms, Darwin's evolution is thus an "open system," non-teleological and non-purposeful — a descriptive mechanism with an undiscovered (or unexisting) outcome. But the direction and outcome of the anthropotropic evolution of technologies seem eminently knowable; indeed, it is the recognition of that obvious direction — namely, that technological media are evolving towards increasing replication of pre-technological communication environments — that serves as the basis of the anthropotropic thesis. Thus the evolution of technological media, from the anthropotropic viewpoint, would seem best explained by a Darwinian mechanism and a non-Darwinian teleology.

Pierre Teilhard de Chardin, whose work will be more fully assessed in the next chapter, has provided just such a model of evolution. Seeking to reconcile the demands of Darwinian evidence with the callings of spirituality, Teilhard suggests that the stuff of the cosmos evolves in a purposeful progression from simple to complex, inorganic to organic, conscious to meta-conscious (the state presently achieved by humanity, the spearhead of Teilhard's evolution), finally culminating in a reconvergence with "that floating, universal entity from which all emerges and into which all falls back,"[8] the "Omega" point of origin and conclusion, or God. The environment, in Teilhard's system, in effect thus selects with a divine sentience — and the variety of mutations from which it chooses are more than merely "accidental" — so that evolution proceeds along its inevitably ascending spiritual plane. Without judging the suitability of Teilhard's model for the universe at large or biological organisms here on Earth, it may be readily applied to the evolution of technologies. As is being argued in the present inquiry, the point of both origin and return for technology is humanity — we are the "Omega/God" of the technological world, the source from which technologies spring and to which they relentlessly return. Moreover, just like the Omega/God in the Teilhardian universe, the human god in the technological universe pervades and informs all aspects of technological evolution: we are literally the "sentient" environment, we are literally the more-than-accidental transmitting agents, and we are literally the terminus of technological media that evolve towards greater convergence with human modes of communication. Thus, a provisional model of anthropotropic evolution would be partially Lamarckian and mostly Darwinian in operation, yet more appropriately Teilhardian (adapted and reduced to human proportions) in direction and destination.

8. *The Phenomenon of Man*, trans. Bernard Wall, with an Introduction by Julian Huxley (original ed., 1955; trans. ed., New York: Harper & Row, 1959), p. 258.

It is precisely this last point — the increasing convergence of technological media towards human systems, which is the heart of the anthropotropic thesis — that is the one glaring omission in Peter Medawar's otherwise full prospectus for a theory of technological evolution. Thus, while Medawar pays special attention to the human role of selecting as well as generating technologies — see his above points (c) and (d), respectively— he seems unaware that the continuing outcome of these processes are technologies which produce environments that are increasingly pre-technological or human. In effect, Medawar performs the highly useful service of describing the tracks, train, and terrain of technological evolution — but neglects, in the end, to tell us in what direction the train is traveling.

LEWIS MUMFORD: THE LOST DISCOVERY

Just the reverse occurs in the work of Lewis Mumford, who notices the anthropotropic tendency of communication media, but only as a seemingly off-hand phenomenon, attributing to it no greater theoretical framework or significance. Thus Mumford, whose recent antipathy to technology has already been alluded to, noted as early as 1934 that

> With the invention of the telegraph a series of inventions began to bridge the gap in time between communication and response despite the handicaps of space: first the telegraph, then the telephone, then the wireless telephone [radio], and finally television. As a result, communication is now on the point of returning, with the aid of mechanical devices, to that instantaneous reaction of person to person with which it began; but the possibilities of this immediate meeting, instead of being limited by time and space, will be limited only by the amount of energy available and the mechanical perfection and accessibility of the apparatus. When the radio telephone is supplemented by television, communication will differ from direct intercourse only to the extent that immediate physical contact will be

impossible. [9]

In this one paragraph written nearly half a century ago, Mumford grasps the full essence and subtlety of anthropotropic evolution, placing media in a clear rank order of development which suggests that, as media have evolved, they have not only extended across space and time, but have extended across space and time in a way that increasingly recaptures the texture of non-extended, or person-to-person communication with which all communication began. Written well before Polaroids, holography, computers, and-transistors, at the dawn of television, color, and synchronized-sound movies, such observation bespeaks enormous insight and prophesy; indeed, had Mumford chosen to investigate the full implications and potentials of his observation, the present study would surely have been unnecessary.

9. *Technics and Civilization* (New York: Harcourt Brace, 1934), cited in Mumford, The Pentagon of Power, p. 295.

Instead, Mumford draws a solitary, disquieting lesson from his rich discovery — namely, that the technological return to pre-technological intimacy poses great injury to rational society. Thus, Mumford contends

> that the maintenance of distance both in time and space was one of the conditions for rational judgement and cooperative intercourse, as against unreflective responses and snap judgement .... The lifting of restrictions upon close human intercourse has been, in its first stages, as dangerous as the flow of populations into new lands: it has increased the areas of friction . . . and has mobilized and hastened mass-reactions, like those which occur on the eve of war.[10]

This problem of "instantaneous access" is a valid one which has turned up in the present study before (see, for example, McLuhan's concern about "discarnate man," in the previous chapter); and while a fuller accounting of it and Mumford's pessimism must await Chapter 9, it can be pointed out now that Mumford's criticism of media evolution for its increasing "immediacy" is jarringly inconsistent with a grander, more frequent indictment that Mumford has hurled against modern technology. In that indictment, elegantly presented in a series of works culminating in 1970 with *The Myth of the Machine, Volume 2*, Mumford attacks technology as "a swollen fragment of man's mind,"[11] which grossly deforms the "organic" biological and life-giving balance from which man arose. "Our situation today calls for a development of the repressed and dwarfed functions of the human personality," [12] Mumford advises. It is perhaps another confirmation of Koestler's observation, then, that great minds are often the last to appreciate the meaning of their own work, that Mumford can be so oblivious to the most profound implication of his own anthropotropic observation: for in making once again possible that "instantaneous reaction of person to person with which communication began"— in retrieving the "repressed and dwarfed functions" of face, color, and sensory nuance — modern communication technologies are reconstituting precisely the kind of organic balance that Mumford calls for. In effect, Mumford focuses only upon the "instantaneous reaction" to the complete exclusion of the growing "person to person" capacity of media; in the end, then, it is Mumford's own one-sided and cursory appraisal of technological media that is the "snap judgement."

10. Ibid,

11. *The Pentagon of Power*, p. 384.

12. *Interpretations and Forecasts* (New York: Harcourt Brace Jovanovich, 1973), p. 290.

Perhaps Mumford also suffers from the same "maiden voyage" syndrome that is the great strength and weakness of McLuhan. Mumford was a pioneer in the charting of "cultural history" — his 1934 *Technics and Civilization* (the source of the anthropotropic observation) is considered by many the first full-scale treatment of how human mastery of symbolic and physical environments has determined the shape and direction of our history.[13] But Mumford covers so much new ground, makes so many previously unnoticed connections and observations, that he is almost bound to miss more than a few of their implications and possibilities. But the irony of Mumford is more than letting a few ramifications slip through his fingers — it is that the magnitude of his subject seems, at last, to have overwhelmed him: for the profound humanist is so busy documenting the impositions of technology that he unwittingly puts humanity in the inferior position. He thus becomes blind to the already-emergent solution to the problem, and becomes part of the very threat he warns against.

13. "Not," as William Kuhns (*The Post-industrial Prophets*, p. 34) has noted, "that Mumford was the earliest; Henry Adams, J. Beckman, and Patrick Geddes had all done important work in the cultural effects of technical innovations. None achieved the scale or synthesis, however, of Mumford's study." For more on the primacy of Mumford's contribution in this area, see Christine L. Nystrom, "Toward a Science of Media Ecology" (Ph.D. dissertation, New York University, 1974); and Peter Haratonik, "Toward the Biotechnic Order: A Study of the Writings of Lewis Mumford" (Ph.D. dissertation, New York University, forthcoming,)

In effect, Mumford sees the direction of media evolution but loses sight of its significance, while Medawar understands the significance of media evolution without ever really determining its direction. The fault in both frameworks perhaps lies in their tendency to address modern technology as a whole, without considering the possible vanguard role of specialized communication technology. (Indeed, with the exception of McLuhan, none of the theorists thus far reviewed accord any special prominence to communication media in their consideration of technology.) But as suggested in the introductory chapter, the anthropotropic phenomenon seems most prevalent in technologies of communication, and may not necessarily be as obvious — or operative — in other technologies, or in technology in general. Thus, Mumford and Medawar may be missing the trees of media for the forest of technology.

In the case of Harold Innis, however, the last scholar to be discussed in the present chapter, communication occupies the back, front, and center stages of a far-flung analysis of technology and history; and while that analysis often seems innocent of both the direction and implications of anthropotropic evolution, it nonetheless offers a provocative and well-developed conception of its own, a conception of media change which holds some interesting leads for anthropotropic theory.

HAROLD INNIS: A DIALECTIC OF MEDIA

If Marshall McLuhan presents media determinism in its most plausible light, then Harold Innis is surely the father of that perspective— for McLuhan himself avers that "I am pleased to think of my own book *The Gutenberg Galaxy* as a footnote to the observations of Innis on the subject of the psychic and social consequences, first of writing and then of printing."[14] But whereas McLuhan's media determinism is fundamentally personal — investigating how technological media alter psycho/ physio sensory ratios — Innis' is more social and political: he is interested in how media alter whole societies and civilizations. Thus Innis, more than any other modern theorist, carries the grand, impersonal, economic determinism of Marx to the study of media: understand how societies transmit and store information, says Innis, and you understand all you need to know about societies, their triumphs and failures, politics and religion, economics and art, the stuff that makes and breaks empires and epochs.

14. McLuhan, Introduction to *The Bias of Communication* by Harold A. Innis (reprint ed., Toronto: Univ. of Toronto Press, 1964), p. ix. Kuhns, p. 140, and Jonathan Miller, *Marshall McLuhan* (New York: Viking, 1971), p. 79, trace Innis' media determinism to the more general technological determinism of Robert Ezra Park, which perhaps in turn may be traced to the economic-technological determinism of Marx, evidenced in such statements as "the hand mill will give you a society with a feudal lord,

the steam engine a society with the industrial capitalist," as cited in David McLellan, *Karl Marx* (New York: Viking, 1975), p. 40. Note that this view of technological communications (Innis) and technological means of production (Marx) as determining or conditioning societies is a good deal less severe than the view, most developed by Ellul but also traceable to Marx, that technologies are becoming our masters. See also above Ch. 2, notes. 16 and 17.

In *Empire and Communications* (1950) and *The Bias of Communication* (1951), appearing a little more than halfway in time between Freud's and McLuhan's main contributions to media theory, Innis sets forth his claim for communications in an ambitious analysis that may be summarized as follows: (a) technological media tend to extend primarily across time or space, but rarely in a balanced combination of both (an interesting refinement of usual extensional theory); (b) the predominating media in most societies tend to congregate in either a time- or space-extending orientation, but rarely in a balanced combination of both; (c) time-extending media tend to encourage societies to be static and tradition-bound, while space-extending media promote expansion and rapid change (a combination of time- and space-oriented media fosters a balance of tradition and change, which Innis regards as healthiest but rare); and (d) media (and consequent societies) with an orientation or "bias" in one direction are usually succeeded by media (and societies) with an orientation towards the other direction: that is, time-oriented media and societies are usually replaced by media and societies that are space-oriented, and vice versa.

15. *Empire and Communications*, revised by Mary Q. Innis, with a Foreword by Marshall McLuhan (original ed., London: Oxford University Press, 1950; rev. ed., Toronto: Univ. of Toronto Press, 1972); *The Bias of Communication*, with an Introduction by Marshall McLuhan (original ed., 1951; reprint ed., Toronto: Univ. of Toronto Press, 1964).

16 "The character of the medium of communication," Innis writes in *The Bias of Communication*, p. 64, "tends to create a bias in civilization favourable to an overemphasis on the time concept or on the space concept and only at rare intervals are the biases offset by the influence of another medium and stability achieved. . . . The power of the oral tradition in Greece which checked the bias of a written medium supported a brief period of cultural activity such as has never been equalled."

Innis documents this analysis with a chronicle of history which may be compressed as follows: (i) Ancient Egyptian and other early civilizations communicated primarily through the spoken word and writing upon immovable objects; information processed in such ways is more easily transmitted from generation to generation across time than across great distances; the resultant societies were therefore hierarchical, religious, tradition-bound, and past-oriented. (ii) Greek culture culminating in Rome relied upon abstract letters written on portable papyrus; such information is readily detachable from its source for movement across vast distances; Rome was the most expansionist civilization of the Ancient World; (and early Greek society, benefiting from a fleeting balance of time- and space-oriented media, produced the outpouring of philosophy and art that became the source for Rome and many succeeding civilizations). (iii) As Rome crumbled, its papyrus was gradually replaced with crumbly parchment, designed primarily for preservation and storage rather than transportation; the Dark and early Middle Ages of Europe were once again tradition-steeped, religious, and looking towards the past. (iv) The introduction of the printing press in the 15th century made information more readily dispersible than ever before; the resulting expansion of European society expressed itself in such movements as the Protestant Reformation, the Scientific Revolution, and the rise of national states, as well as a

general lack of concern for the past that typifies modern existence, (v) The "new" medium of the radio, barely touched upon by Innis, is said to engender an interest in "stability," and "an increasing concern with problems of time." [17]

17. *Empire and Communications*, p. 170. McLuhan, in his introduction to Innis' *Bias of Communication*, quite properly takes Innis to task here for failing to appreciate that electronic media "in effect abolish [i.e., extend across] space and time alike" (p. xiii). On occasion, however, Innis did recognize the powerful spatial extension of radio, pointing out that it "provided a new base for the exploitation of nationalism and a far more effective device for appealing to larger numbers," *Bias of Communication*, p. 81. Perhaps Innis' confusion in this area reflects a partial awareness that electronic media increasingly extend across both time and space; see the end of this section.

The historical accuracy of Innis' media determinism (e.g., how much of Rome's expansion was the consequence, and how much the cause, of its widespread use of papyrus?) is of less interest to a theory of media evolution than the pattern or mechanism of media change that Innis' analysis uncovers: namely, that media, both individually and collectively, seem to extend in a manner or direction opposite to that of the media that immediately precede them. Although Innis leans heavily on the interplay of time and space extension to support this observation, he sees the alternating pattern in other dimensions of media extension, suggesting, for example, that

> the monopoly of communication based on the eye hastened the development of a competitive type of communication based on the ear, in the radio and in the linking of sound to the cinema and to television. Printed material gave way in effectiveness to the broadcast and the loud speaker. [18]

In other words, visual extension invites aural extension, in Innis' system, in much the same way that an environment of spatial extension ultimately engenders new media that swing in the "other" direction towards time extension (and vice versa). (Note, however, that Innis' contrasting or "competitive" modes of extension are not "opposites" in the mutually exclusive sense that positive electricity and negative electricity, or even capitalism and collectivism, for example, are thought to be. Indeed, rather than necessarily excluding each other, such alternatives as time and space extension are viewed by Innis as capable of complementing each other in balanced media systems, which, as suggested above, Innis views as rare but nonetheless possible and desirable. See discussion of Claude Levi-Strauss, below, for more on complementary opposites in human systems.)

18. *Bias of Communication*, p. 81. It should be noted, however, that Innis often equates "ear" with time extension, and "eye" with space extension, thus making these terms merely sensory designates for the prevailing factors of time and space. The equation of "ear" with time extension, incidentally, may also help explain why Innis tended to view radio as time-oriented, see note 17 above in this chapter.

For all its historical breadth and flair, however, Innis' succession of opposites remains in the end a static mechanism, that goes nowhere. Whether the media are print and radio of the twentieth century, or clay and papyrus of much earlier, the objects of Innis' analysis all seem to be caught in the same interminable ping-pong game that hits them back to time extension, forth to space extension, now back to time extension, and so on, with never an overriding direction, never a resolution, never an evolution. The fixed base of Innis' fluctuating system is aptly captured by William Kuhns, who calls it a "seesaw" and a "pendulum," and then writes that "no one knows what makes the pendulum swing."[19] It turns out, however, that the invisible hand remains invisible only as long as the process of media change is regarded as a pendulum: for once the time and space fluctuations are recognized as an evolving rather than a fixed process, and the direction of that evolution is noticed, the force that guides the fluctuations becomes immediately apparent.

19. *The Post-Industrial Prophets*, pp. 146, 176.

Central to Innis' fixed system— and, indeed, to the pendulum model itself— is a concept of balance which, appropriate to a pendulum, is only fleetingly attained as media swing from one type of extension to the other (to hover around a midpoint— i.e., a balance of space and time extension— is to cease to be a pendulum). Indeed, a strain of pessimism underlies Innis' otherwise detached analysis, a pessimism which stems from Innis' view that desirable balances of space and time extension seem inevitably to give way to one-sided, biased media environments. ("Each civilization has its own methods of suicide," Innis writes.)[20] Such a view, however, fails to take note of a gradual accumulation of space and time balance in modern technologies, culminating in devices such as the computer, which can disperse information almost as easily as it can store information. Even print, considered by Innis the space-extending medium par excellence, has certainly committed more information to extension across time than wall-carvings, parchment, speech and probably all previous time-extending media put together. Indeed, as will be seen in greater detail in Chapters 5 and 6 below, the whole emergence of advanced technologies in the 19th and 20th centuries seems a carefully orchestrated series of ever finer accommodations between contrasting types of extension — with the phonograph in 1877, for example, providing an extension of voice across time which complemented the extension of voice across space achieved by the telephone a mere year earlier, and both phonograph and. telephone providing an acoustic balance to the visual extension achieved by photography and telegraphy some 30 years earlier. This evidence suggests that if media change.is a pendulum, at all, it is an evolving pendulum whose swings from one extension to another grow both shorter and quicker— that is, each swing, whether to time or space (or ear or eye, etc.), carries with it an accumulation.of all previously attained balances, which works to reduce both the extremity and duration of any swing to a one-sided extension. The result is a pendulum that increasingly hovers around a center point of balanced extension (in other words a pendulum, as suggested above, that ceases to be a pendulum).

20. *Bias of Communication*, p. 141.

Now there is nothing in the concept of an autonomous pendulum which would explain why the pendulum should so factor itself out of existence; but once the pendulum is seen as guided by an anthropotropic hand, the momentum towards increasing balance becomes easily explainable and expected. A pre-technological balance of human senses, acting as a determinant of media evolution, has already been suggested in the previous chapter's discussion of McLuhan's "sensory ratios"; in addition to this balance, one might plausibly suggest a balance of time and space perception in human systems, which advancing technologies also increasingly strive to recapture. For just as we usually see and hear rather than see or hear in pre-technological environments, so we usually see and hear in a limited, pre-technological way both through space and time (space perception operating through our external senses of sight, hearing, smell, etc.; time perception operating in our "inner" senses of retrospection and anticipation). Moreover, as would be expected on the basis of anthropotropic theory, primitive "Stage B" technologies break out of pre-technological boundaries only in a gross, unbalanced, "either/or" way— that is, they extend uni-dimensionally, across either time or space (or ear or eye)., but hardly ever, as Innis has correctly observed, across a balanced combination of both. It is only with the increasing evolution of media towards "Stage C," apparently undetected by Innis, that time and space extension once again co-exist in media, in an increasing recreation of the balanced pre-technological environment. Innis' succession of opposites thus becomes a dialectic evolution of media, with syntheses that come closer and closer to the "Absolute Spirit" of human communication. (Indeed, the general anthropotropic model of media evolution seems to fall into a dialectic pattern, with the Stage A "thesis" of unextended, natural communication succeeded by the Stage B "antithesis" of extended, distorted communication, which in turn engenders the Stage C "synthesis" of communication that is both extended and increasingly natural. See Ch. 4, n. 20 below for further discussion of the relevance of the dialectic to the anthropotropic model.)

In failing to notice the increasing time and space (and sensory) synthesis of modern media, Innis joins most other extant media theorists in what may be termed the "Ellulian error" of assessing all of technology by its infancy, or projecting the limitations of the technological caterpillar onto the butterfly. Nonetheless, Innis deserves credit for recognizing what seems a fundamental mechanism of media change which, when coupled with anthropotropic theory, helps to greatly specify the process of media evolution — to wit, that media evolve towards replication of human communication environments not in a deliberate straight line, but in a ping-pong series of over- and under-approximations, a process that vibrates to and fro and only eventually settles upon the human mark. Such a process, moreover, is entirely compatible with the Darwinian method of evolution, previously discussed, in which organisms evolve in a series of inexact trial-and-error accommodations to the natural environment.

Moreover, Innis' observations help develop anthropotropic theory in a second important way: in tracing the succession of media by their "opposites" back to the beginnings of technology, Innis in effect demonstrates the workings of anthropotropic forces, albeit unrefined, at the very earliest stages of technology. For in the anthropotropic perspective, the broadly primitive pendulum so well described by Innis can be seen as an attempt to coarsely compensate for an imbalance in one extensional direction with an equivalent imbalance in another direction. (That the pendulum is powered by an active human hand becomes obvious, as indicated above, with the eventual lessening of the swings in "Stage C" to human proportions.)

Innis' findings thus suggest an initial refinement of the anthropotropic model along the following lines: (1) anthropotropic tendencies — i.e., the propensity of media to replicate human communication environments — operate at all stages of technology, primitive as well as advanced; (2) in the inefficient, primitive technologies of "Stage B," the anthropotropic tendency is manifested in the replacement of one uni-extensional distortive technology with another, an attempt to "rectify" one gross imbalance with an "opposite" imbalance; the net result of this "compensation" is of course usually just another imbalance, so that the anthropotropic movement in "Stage B" can be said to be "unsuccessful;" (3) only with increasing technological sophistication does the compensation process become refined enough to achieve a genuine extensional balance; this gradual retrieval of pre-technological, multi-dimensional communication is the fulfillment of the anthropotropic process, still emergent in current "Stage C" technology. The difference between early and advanced technologies thus becomes not so much the absence of anthropotropic evolution in the first and the presence of anthropotropic evolution in the second, but rather the failure of anthropotropic evolution in the first and its success in the second.

In addition to confirming the applicability of the Darwinian trial-and-error model for media, and lengthening the history of anthropotropic operations, Innis performs yet a third, though more obvious, service for the present inquiry. In writing of societies that are generally "time-oriented," "space-oriented," and so forth, Innis emphasizes the tendency of disparate media to have a unified, rather than individual or separate, impact upon culture and society. Such conjunctive workings raise an important question for anthropotropic theory: to what extent do media evolve individually towards replication of human environments, i.e., to what extent does each medium attempt to replicate the full gamut of the human environment; and to what extent do media evolve towards replication of human environments in an interrelated way, with each medium replicating only an interlocking piece of the human jigsaw? Does it make sense, from the anthropotropic viewpoint, to consider the evolution of radio without recording technology, or the development of television without video-tape? Biological models of evolution increasingly speak of "co-evolution" of symbiotic organisms;[21] and since biological models, as indicated above, hold many important lessons for media, the possibility of a "co-evolution" of media will receive close attention in the ensuing study (see Chapter 6 below).

21. A mailed advertisement for *The CoEvolution Quarterly* (Box 428, Sausalito, California) explains that the term was introduced in 1965 by Paul Ehrlich and Peter Raven to suggest that "relationships" among organisms rather than organisms per se are what evolve.

Thus Innis, himself uninterested in a purposeful, general theory of media evolution, makes in many ways the most concrete contributions to an anthropotropic model. As for the validity of the sweeping media determinism that is his central thesis — the question of to what degree the fluctuating societies are the product and to what degree the cause of the fluctuating media — this is an intricate problem, already alluded to in connection with McLuhan above, and to be briefly discussed at various junctures in the study below.

* * *

In sum, then, the existing theoretical leads on media evolution come from three sources: (a) Medawar's suggestion that technological extensions evolve much like biological organs and organisms, and his brief exploration of Lamarckian and Darwinian analogies; (b) Mumford's explicit recognition of anthropotropic evolution, diminished by his unfortunate failure to appreciate its implications; and (c) Innis' discovery of a prime mechanism of media evolution — a Darwinian-like ping-pong process of approximations — and his documentation of the anthropotropic process back to the beginnings of technology. All three may be considered incomplete, but important, preformulations of anthropotropic theory.

CHAPTER 4

ANALOGUES AND PARTICULARS

Nearly all the theoretical viewpoints thus far considered have in some way or another directly concerned technology. Even theorists such as Freud may be considered primary sources for the present study, since they make observations that are explicitly technological, albeit briefly. There exists, however, a varied body of work concerning human transformations and extensions that are not "technological" in the usually defined sense; and there exist numerous models of evolution for phenomena other than media. The theories of Lamarck and Darwin, discussed above in Chapter 2, are examples of the second kind; Noam Chomsky's treatment of language as a series of "transformations," to be discussed below, is an example of the first. From the technological perspective, these theoretical systems are valuable metaphors or "analogues": they may contain useful insights not present in the thin ranks of primary technological theory. At the same time, however, the demonstrably different subject matter of the analogues requires that their applicability to technology be carefully weighed, with points of relevance and non-relevance distinguished as best as possible.

At the other end of the gamut, so to speak, are several communication perspectives which, differing in another way from the theories thus far considered, tend to specialize in a particular medium or communication process. Various "schools" of film criticism and theory, for example, have for the most part been primarily concerned with a single medium — film— and, with the exception of tangential references to photography and theater, have rarely addressed the operation and evolution of other media, or of communication technology in general. Elements of some of these film analyses, however, seem easily generalizable to all technological media, and thus hold some value for anthropotropic theory. Cybernetics is another discipline that tends to draw examples from one technology, computers— yet here, as will be seen below, the lessons are not only generalizable to all communication media, but suggest a fundamental, physical reason for the very process of anthropotropic evolution itself. Such "specific" communication models will be considered after the analogues.

NOAM CHOMSKY: DEEP STRUCTURES AND TRANSFORMATIONS

As suggested in Chapter 2 above, the concept of technology as an extension of human systems is rooted in an older perspective, perhaps first developed by Marx, that views all human activities and artifacts as extensions or "transformations" of a few underlying, often genetically inherent and unconscious/human mental characteristics or "structures." In addition to its application, starting with Freud, to problems of technology, such transformationalism or "structuralism" has also been recently employed in analyses of other media-like phenomena, most notably in Noam Chomsky's model of language and Claude Levi-Strauss' appraisal of myths.[1] Indeed, the structuralist approach in both these disciplines has been far more developed than any possible counterpart in media theory, attempting in many instances to provide a rather complex mapping of the ways that genetic mental traits become transformed into observable human activity. A brief examination, then, of some of the non-technological transformational models should offer some additional clues about how technology extends and reflects fundamental human processes.

1. Use of the term "structuralism" to describe Chomsky's perspective on language may be a bit misleading, since the older, more static approach to linguistics that Chomsky's transformational model has replaced is also referred to as "structural linguistics." However, in the sense that the term "structuralism" is now most typically used — to describe those perspectives that consider human activities as a series of transformations of inherent mental structures — the inclusion of Chomsky as a structuralist along with Levi-Strauss and Freud would seem entirely warranted. See Howard Gardner, *The Quest for Mind* (New York: Knopf, 1973), pp. 241-243, for a comparison of Levi-Strauss and Chomsky, and a discussion of Chomsky as a "structuralist"; see Richard DeGeorge and Fernande DeGeorge, eds. *The Structuralists from Marx to Levi-Strauss* (New York: Anchor, 1972), pp. xi-xxix, for a brief discussion of the general structuralist approach, along with summaries of various structuralist models; see John Searle, "Chomsky's Revolution in Linguistics," in *On Noam Chomsky*, ed. Gilbert Harmon (New York: Anchor, 1974), pp. 2-33 for a comparison of Chomsky's work and the earlier, non-transformational "structural linguistics" that Chomsky's approach has replaced.

Just as media theorists are increasingly interested in the organization of our information environment rather than its mere content, so Noam Chomsky has been interested in the organization of language — the "rules" which govern its practice, i.e., its "grammar" or syntax— more than its semantic content. In numerous publications throughout the past twenty years, Chomsky has developed a transformational account of the operation of such language rules or grammar, which may be summarized as follows:[2] (a) from the grammatical perspective, the smallest unit of speech is not a word but a phrase or "structure"; thus, "I go" is an expression of a fundamental grammatical unit, whereas "I" or "go" without the implied verb or subject is grammatically indecipherable; (another way of explaining this is to point out that "I go" bespeaks an attitude or process

— in this case, one of "positiveness"— whereas "I" or "go" by itself does not) ; (b) most grammatical units of speech are transformations of more primary or "deeper" grammatical units; language can thus in theory be analyzed by finding its initial or deepest grammatical units, and tracing the number and types of transformations the units have undergone; e.g., "I can't go" may be considered a negative, conditional transformation of the more fundamental "I go" unit; (c) the deepest or most fundamental grammatical units are genetically embedded in the human brain; that is, every mentally undamaged human being has the ability to generate a finite number of grammatical structures or relationships, which, when appropriately transformed, may be potentially understood by other humans; (d) the transformational possibilities are themselves genetically prescribed; that is, there is a finite variety of transformations that any grammatical structure can undergo. In view of such an analysis, the task of the transformational linguist would lie in identifying the earliest — i.e., genetic — grammatical structures, and in discovering and detailing the transformational rules by which the genetic structures are converted into the almost infinite possibilities of actual language usage.[3] Methods thus far used for such a task have included cross-linguistic comparisons (discovery of grammatical similarities in disparate languages would point to common genetic structures) and monitoring of the language acquisition process in children (discovery of certain grammatical mistakes that children never make would also suggest inherent grammatical patterns).

2. See Gilbert Harmon, *On Noam Chomsky*, and Chomsky's own *Reflections On Language* (New York: Pantheon, 1975) for the most up-to-date presentations of Chomsky's thinking on language. See also Paul Levinson, review of *Reflections On Language* by Noam Chomsky, in *Media Ecology Review* 4 (May 1976): 24-26.

3. Recently, however, Chomsky, pp. 25, 124 has suggested that some mechanisms of language usage may be forever beyond "scientific" understanding. In effect, Chomsky parallels an argument made by Gunther Stent in "Limits to the Scientific Understanding of Man," *Science*, 21 March 1975, pp. 1052-1057, that scientific knowledge (i.e., knowledge that is logically or empirically testable) is a biologically adaptive device subject to the rigors of natural selection, and hence only obtainable in situations where such knowledge would be of demonstrable assistance to the survival of the human species. This limits scientific inquiry to the external Newtonian world (in response to which our species evolved), and excludes both the internal world of human grammatical constructs (which was not part of our natural environment) and the cosmic universe addressed by Einsteinian physics. At best, such a contention is contradicted by the palpable fact that Einstein was able to produce a theory of the universe that is at least in part scientifically testable and understandable by other human minds; at worst, the "biological limitation" hypothesis in Chomsky's case is easily mistakable as a rationalization for failure to thus far uncover many of the elements of human mental organization. A more detailed criticism is planned by the present author in a forthcoming article; see also the end of Chapter 6 below for consideration of language as a "natural" environment.

To what extent are the methods and goals of transformational linguistics transferable to a theory of anthropotropic media? Chomsky himself has been ambiguous as to the wider applicability of his model, writing first, in *Language and Mind* , that "in general, the problem of extending concepts of linguistic structure to other cognitive systems seems to me, for the moment, in not too promising a state";[4] but changing his mind, some seven years later in *Reflections On Language*, where he suggests that in the same way that transformational grammar is studied,

> **one may also want to isolate for special study the faculties involved in problem solving, construction of scientific knowledge., artistic**

creation and expression, play, or whatever prove to be appropriate categories for the study of cognitive capacity, and derivatively, human action. [5]

Although Chomsky's about-face seems contradictory, it seems possible that he may have been right both times.

4. 2nd ed. (original ed., 1968; rev. ed., New York: Harcourt Brace Jovanovich,- 1972), p. 75.

5. *Reflections*, p. 35.

The strengths and weaknesses in the application of language constructs to the understanding of media are in many ways typified by the applicability of the key linguistic concept of "deep" or "initial" structures. These primary grammatical units seem to coincide in many ways with the anthropotropic description of "Stage A" pretechnology or human communication forms: both operate at the origins of their systems, and to varying extents determine, and are reflected in, the ultimate disposition of those systems. Examples of such "deep" media or communication structures range from the obvious perception of color and depth in the real world, to the more complex organizational constructs of simultaneity, balance of senses, and balance of time and space perception in human systems described earlier in the present study. It may thus be possible to inquire of technological media as to what are their deep human structures — i.e., what pre-technological communication forms do they embody or represent— and to gauge, as will be attempted later in this inquiry, the survivability of any medium on the basis of how fully, or with what fidelity, it replicates or extends its deep or "Stage A" antecedents.

But certain shortcomings in the linguistic/media analogy immediately suggest themselves. In the first place, while deep media structures, like their linguistic analogues, are not always observable in their own technological systems (i.e., you may not be able to identify the deep structures of television just by observing television), they are, in marked contrast to their linguistic cousins, abundantly in evidence in the larger human system. To identify the pre- or non-technological forms of communication, it is necessary merely to consult relatively recent history, or the non-technological situations that persist in many underdeveloped parts of the world, and indeed even in advanced technological societies such as our own (e.g., despite the pervasiveness of two-dimensional television, we still spend a large sector of our lives seeing in three dimensions in the real world — which allows identification of the third dimension as a deep visual structure, imperfectly transformed by two- dimensional television). This accessibility of deep media structures means that many of the methodologies of transformational grammar, geared towards illumination of hidden or otherwise recondite initial language patterns, are unnecessary or inappropriate for the study of media.

A second point of divergence between the deep structures of language and technological media lies in the way that the deep structures tend to "determine" or condition the surface manifestations in each of their respective systems: while the deep structures of grammar serve to determine actual speech only in the sense that they limit or prescribe ultimate speech possibilities, the deep structures of media serve, as has already been indicated in the present study, as literal models which technologies seem increasingly to replicate. Thus, actual language often bears no obvious resemblance to its originating structures (which is why deep language structures are often so difficult to isolate and identify), while technological media resemble their human antecedents more and more, apparently moving towards an ultimately complete convergence of sorts. It follows, of course, that the "transformational rules" of each system would operate in a commensurately different manner — the transformations of language being genetically fixed or immutable (the diversity in surface language deriving from the huge number of permutation possibilities that arise in applying a fixed repertoire of transformations to a fixed repertoire of deep structures, rather than variability in the deep structures or transformations themselves), while the transformations of media continuously evolve towards more accurate replication of deep human structures. Indeed, in the terminology of deep structural/transformational grammar, the present study of anthropotropic media can be characterized as a study of how media transformations are becoming increasingly "perfected," so as to provide closer and closer approximations of the pre-technological deep structures of human communication.

The question arises as to whether the deep structures of communication are genetically inherent in humans, as the deep structures of grammar are thought to be. It is Chomsky's insistence that the initial patterns and transformational rules of language are built into the species, of course, that has been the most provocative element of his model — leading to a view that numerous human "cognitive capacities" are, like language, genetically derived; and pointing, in general, to a serious reconsideration of the Cartesian thesis of innate mentality. Yet as Edward O. Wilson has pointed out in *Sociobiology*,[5] the deep structural/transformational model of grammar seems capable of working without recourse to genetics: it is possible, for example, that initial language structures and rules of transformation are unconsciously acquired early in childhood, and thereafter exert the type of determining effect upon actual speech that Chomsky and others have so well documented. Such a hypothesis would explain the difficulty encountered thus far in uncovering components of a "universal" genetic grammar (and would, for the same reasons, tend to be falsified by such similarities in diverse languages as have been identified).

5. Pp. 558-559.

For the introductory purposes of the present inquiry, at least, it seems that the anthropotropic theory of media too can work without postulating a deep structure — or a pre-technological regimen — of communication that is necessarily the product of genetics. Stated in its simplest form, the anthropotropic thesis holds that technological media increasingly perform like pre- technological or non-technological modes of communication — that the pre-technological deep structures, in Chomskyan terms, "determine" the surface technologies; or, more colloquially, that "natural" communication is increasingly emulated by "artificial" communication. Support for such a view — to be presented in subsequent chapters — lies in examples of technological communication that increasingly resemble communication either without or before technology. While recognition of such resemblances of course requires prior identification of pre-technological or deep communication structures, there seems no immediate need to know the origin or the source of the pre-technological structures themselves. Thus, the anthropotropic thesis is sustained, for example, when we notice that (1) color perception exists both before and without technology and (2) technologies have evolved from no-color to fuller color replication. Whether human color perception itself is genetically instilled, socially acquired, or some combination of both is for the above purposes not pressingly relevant. In general, then, the derivation of pre-technological structures — other than their placement either before or without technological media — need not occupy a large portion of an initial anthropotropic inquiry, and will thus remain, along with other intriguing problems listed at the close of the introductory chapter, for the most part unexplored in the present work.

It must be noted in passing, however, that to the extent that the anthropotropic model holds as an accurate description of events, it strongly implies a genetic basis of pre-technological structures. In the first place, the very stability and longevity of pre-technological patterns, which not only survive but serve as magnets for technological evolution, suggests a genetic rather than environmental or cultural origin. Why, if pre-technological modes of communication were merely products of an earlier culture, would they withstand the considerable environmental pressure of subsequent technologies so thoroughly gauged by media determinists? (Of course, it is altogether possible that even genetic patterns of communication might change in response to selective pressures from media environments of several thousand years or more, such as abstract writing — a likelihood that will be further explored in Chapter 6 below.) And why, if pre-technological communication patterns were primarily social or conventional in origin, should such initial social patterns dictate the performance of future social technologies, and seem increasingly to function as explicit models for those technologies? Surely the fact that the non-technological patterns came first is not enough, since, as the present study has already indicated, there have been primitive, i.e., Stage B, technologies, subsequent to pre-technologies yet prior to advanced technologies, which have hardly served as models for newer technologies and indeed have often been "contradicted" by them, as in the case of black-and-white replaced by color photography and silent movies superseded by talkies. The explanation for such phenomena may well reside in the special, i.e., genetic or non-social, source of pre-technological communication structures: it is possible that every human generation, regardless of its specific media environment, has a built-in "sense" of what communication should be like — a sense that makes itself felt both in invention and consumer selection of communication technologies, and hence in their eventual evolution. Further support for the innate origin of deep communication structures comes from the biological nature of many pre-technological patterns already identified in the present study: much long-standing evidence suggests that depth perception, for example, is more physiologically than culturally determined,[7] and the role of balance or equilibrium has long been recognized as a key feature of the biological and naturally physical worlds. Thus, while the notion of genetic pre-technological structures is by no means necessary for an anthropotropic theory of media evolution, genetic origins seem a reasonable explanation for much of the evidence accumulated in an anthropotropic perspective; as anthropotropic theory is pursued and developed, it may therefore be expected to become generally aligned with disciplines such as sociobiology, Gestalt psychology, and various "structuralisms" such as Chomsky's, which argue for a genetic derivation of much of human activity.

7. The "visual cliff" demonstration, which shows newborn animals and humans responding to depth cues, is the classic in this area. See Ernest Hilgard, *Introduction to Psychology*, 3rd ed. (New York: Harcourt, Brace & World, 1962), pp. 208-210.

Thus, Chomsky's transformational model of grammar is a valuable stimulant for anthropotropic theory — not only in opening bridges to another system, but in forcing anthropotropic theory to carefully examine the places where the two systems seem to fail to coincide. The result of this process has been a shrinking and stretching, and general reappraisal, of transformational/structuralist terms and concepts for suitability to anthropotropic theory — and a growth of anthropotropic theory itself. Having thus "teethed" on Chomskyan structuralism, however, anthropotropic theory can expect to receive only diminishing returns from lengthy comparisons with other structuralist disciplines — such exercises presumably revealing once again, albeit from different examples, that media transformations are evolving rather than fixed, media deep structures are accessible not hidden, and so forth. There is, nonetheless, at least one element in the structuralism of Claude Levi-Strauss which bears special attention. [8]

8. See Edmund Leach, *Claude Levi-Strauss*, rev. ed. (New York; Viking, 1974) for a summary of Levi-Strauss' work.

CLAUDE LEVI-STRAUSS: BI-POLAR MENTALITIES

Claude Levi-Strauss, a cultural anthropologist, proposes a broad structuralism that encompasses the gamut of human activities; in situations ranging from the abstractions of religion to the fundamentals of food preparation, Levi-Strauss deduces a common set of mental "deep structural" characteristics which organize all our cultural affairs. Perhaps foremost among these mental matrices is the principle of "bi-polar opposition," the tendency of the human mind to fragment the continuous environment and then arrange the parts in "opposite" pairs, such as male and female into "opposite" sexes, heaven and hell as opposite alternatives for afterlives, and even red and green as "stop" and "go" in traffic signals. Although some opposites do seem already present in nature — for example, black (no color) and white (all colors)— structural anthropologists have noted that in cases such as those listed above, the opposition exists not in the original phenomena, but in the perception and ordering of them in the human mind.

Such a process might explain the tendency of media, noted by Innis and already discussed above, to develop in a "succession of opposites": as extensions and artifacts of the human mind, technological media would seem plausibly subject to the same mental dictates that polarize cultural perceptions and thoughts. In effect, this application of the "bi-polar opposition" principle to media entails only the small step of suggesting that human beings, in addition to thinking in opposites, also construct in opposites. Moreover, the very way that the Levi-Straussian system defines "opposition" dovetails with the peculiar way that media seem to function as "opposites" — that is, time extension is opposite to space extension, and ear extension opposite to eye extension, only in the limited, relative sense that red is the opposite of green, or male the opposite of female. Of course, differences in the nature of conception (thinking) and technology (making) account for differences in the operation of binary opposites in the two systems. Thought, as the more immediate and flexible product of the mind, allows for simultaneous expression of opposites (both poles must be already present in the deep structural mind, hence simultaneously present in the mind, if they are to be expressed in culture); whereas in more cumbersome technology, opposites tend to appear in succession, as Innis has documented. Yet if anthropotropic expectations are correct— if technologies, in structuralist terms, grow ever more congruent to deep communication structures— then technologies should increasingly operate in contrasting ways at the same time. And this, as has been already noted at the end of the previous chapter, indeed seems to be the case: not only do modern technologies, viewed as a whole, increasingly extend across time and space, ear and eye, but individual communications media such as motion pictures, television, and holography seem to increasingly combine contrasting sensory extensions (see also "Co-evolution and Convergence" in Ch. 6 below). Thus , Levi-Strauss' principle of bi-polar opposition not only helps explain the succession of opposites of primitive media noticed by Innis, but fits comfortably into the additional anthropotropic observation that media are evolving towards greater balance, or inclusion of contrasting traits in the same operation. (See also earlier discussion of McLuhan's view that evolution from print to electronics is evolution of uni-extensional to multi-extensional media, at the end of Chapter 2.)

Ironically, it is on this very issue of evolution or change that the various "schools" of structuralism, including Chomsky's and Levi-Strauss', have least to contribute. Although the structuralists view language, culture, and human existence itself as a series of transformations, the transformations themselves are seen as permanent and unchanging, a genetically fixated set of human possibilities; indeed, in their single-minded search for constants across continents and history, these theorists often seem close to embracing an Aristotelian, static view of the universe. Structuralism thus provides analogues appropriate to technological transformation, but inappropriate for technological evolution. For models of evolution, the anthropotropic theory must turn elsewhere.

The Darwinian, Lamarckian, and Teilhardian models of biological (and, in Teilhard's case, cosmic) evolution, previously discussed, provide the fundamental analogues for anthropotropic evolution — suggesting, in bits and pieces, the various intricate mechanisms by which technologies achieve consonance with the pre-technological world. But are there, in the absence of applicable structuralisms, any models for the evolution of thought processes and ideas, from which an anthropotropic model of technological media evolution can gain insights? Since at least the time of Hegel, conceptions of the evolution of culture and ideas have been numerous; but the theory of the evolution of ideas that holds the most possibilities for anthropotropic theory is the one recently proposed by Karl Popper.

KARL POPPER: A DARWINIAN EVOLUTION OF IDEAS

Popper's view that scientific knowledge accrues from attempts to "falsify" or confront all assumptions has already been cited as a philosophical underpinning of the present study. In *Objective Knowledge*, Popper pursues the problems of knowledge and reality further, and suggests that "reality" or existence comes in three interrelated forms: "World One," consisting of physical objects and states; "World Two," the realm of subjective or immediate mentality, as in feeling or the act of thinking; and "World Three," the products of human mentalities, the collective body of human knowledge which once produced is semi-autonomous or "objective," and retrievable not only in individual memory but in books, computers, and the variety of technological recording apparatuses. [9]

9. A more recent and more fully developed description of World Three is provided by Popper in *The Self and Its Brain*, p. 38: "By World 3, I mean the world of the products of the human mind, such as stories, explanatory myths, tools, scientific theories (whether true or false), scientific problems, social institutions, and works of art." In emphasizing that material as well as nonmaterial products of the mind are World Three constituents, Popper expands that world to include not only objective "knowledge," but all cultural artifacts. Popper's World Three thus begins to offer the first "philosophic" treatment of modern technology — that

is, a consideration of technology in the context of such traditional philosophic problems as mind/body, reality and existence, origins of knowledge, and so forth (as distinct from the historical approaches, for example, of Mumford and Innis, or the literary/poetic method of McLuhan). (See also Ch. 6, n. 12 below.)

It is the realm of World Three which, as the realm of human knowledge, most interests Popper; and it is the realm of World Three, which as will soon be seen below also encompasses technologies, that most interests the anthropotropic model.

For Popper proposes that World Three, sired by World Two and residing in World One, possesses a biological-like life— and evolution— of its own:

> ... the growth of objective knowledge ... can be interpreted as a description of biological evolution....
>
> The tentative solutions which animals and plants incorporate into their anatomy and their behaviour are biological analogues of theories; and vice versa: theories correspond (as do many exosomatic products such as honeycombs, and especially exosomatic tools, such as spiders' webs) to endosomatic organs and their ways of functioning. Just like theories, organs and their functions are tentative adaptations to the world we live in....[10]

Theories and knowledge, in other words, are biological"adaptations" to the surrounding world— strategies for dealing with "reality," which live or die depending upon how well they approximate or respond to the "truth" of their environments. Popper is thus led to see the evolution of these theories in explicitly Darwinian, or trial-and-error, terms:

> The growth of knowledge — or the learning process — is not a repetitive or cumulative process but one of error-elimination. It is Darwinian selection, rather than Lamarckian instruction.[11]

10. *Objective Knowledge*, p. 145.

11. Ibid., p. 144.

Although Popper never fully addresses the question of where human communication technologies fit in this schema, it is fairly obvious that technologies, too, are human theories or strategies for dealing with the outside world. Technologies, like ideas, after all, are game-plans that survive based upon how accurately they incorporate or reflect the truth of the situations they may function in (anthropotropic theory suggesting that the "truth" against which communication technologies are measured is their ability to extend across space and time without disrupting the pre- technological environment). Moreover, communication technologies in particular appear to have a special World Three character: while all technologies are both World One physical entities and embodiments of World Three ideas or products of the human mind (e.g., a lawn-mower is both a World One metal object and a World Three human strategy for mowing the lawn), communication technologies in addition have World Three "contents," which usually consist of ideas distinct from those embodied in the technology itself. A book or a radio, in other words, is not only an idea about how to communicate (i.e., a strategy or way of communicating), but obviously a transmitter of other ideas as well (i.e., what is written in the book or said on the radio). (Note that language exhibits the same "double" World Three property as communication technologies — that is, a word or a language is in itself an idea or strategy for communication, as well as a presenter of ideas.)[12]

12. Synonyms such as "demi" and "semi," for example, are two distinguishable strategies or ways of communicating the identical concept of "halfness," in much the same way as books and radio, for example, are two different ways of communicating the word "tree." In this sense, communication technologies may be more properly considered "languages" than "tools" — a view expressed some twenty years ago by Edmund Carpenter in "The New Languages," reprinted in *Explorations in Communication*, eds. Carpenter and Marshall McLuhan (Boston: Beacon, 1960), p. 162, to wit: "All languages are mass media. The new mass media — film, radio, TV — are new languages." See also n. 17 below in this chapter, and "The Persistence of Abstraction" in Ch. 6 below.

Popper appears to recognize the special World Three character of media, and its link with language, when he writes that

> the kind of extra-personal or exosomatic evolution that interests me here is this: instead of growing better memories and brains, we grow paper, pens, typewriters, dictaphones, the printing press, and libraries. These add to our language . . . what may be described as new dimensions. The latest development . . . is the growth of computers.[13]

But Popper pursues the matter no further, and indeed seems to misconstrue the media-World Three relationship in two ways: (a) as the above excerpt suggests, he apparently limits the language- or brain-extending media to those involved in writing or print (plus the computer), overlooking the fact that when a medium records or transmits speech (which telephone, television, and virtually all media, with the exception of still photography, in some way or another now do), the medium is per force also extending language; (b) he subsequently contends that while "our tools and instruments" (presumably including communication technologies) seem to evolve from general to more differentiated forms (an example of this might be the evolution of fire as a general cooking technology into ovens, broilers, toasters, etc.), pure knowledge "grows almost in the opposite direction . . . towards increasing integration towards unified theories" (e.g., Einstein explains more than Newton).[14] Aside from many indications that Popper's assertion about the evolution of tools is itself not entirely accurate (modern cooking technologies, for example, actually integrate many previously specialized cooking functions), [15] the generalized- to-specialized pattern seems wholly inapplicable to modern communication technologies, which, as will be described more fully in Part II below, seem rather to be moving in the fragmentary-to-unified direction that Popper discerns in the evolution of theories and knowledge. For example, "talking" motion pictures in effect integrated the silent movie with the phonograph, television combined the content of movies with the delivery system of radio, and future television seems likely to subsume both telephone and newspapers. [16] Thus, communication technologies seem not only analogous to ideas in that both evolve in a Darwinian-like process of selection, but share a further identity with ideas — perhaps arising from a similarity in the underlying purposes of communication and ideation, both of which attempt to "bridge" different situations — in that both seem to evolve in a predominantly integrating, or unifying pattern. [17]

13. *Objective Knowledge*, pp. 238-239

14. Ibid., p. 262.

15. See, for example, *Consumer Reports Buying Guide Issue 1978* (Mt. Vernon, N.Y.: Consumers Union, 1977), p. 18, for a description of "toaster oven/broilers." Popper (*Objective Knowledge*, p. 262) cites Herbert

Spencer's view that organisms evolve towards greater differentiation as a model for Popper's view of the evolution of tools. Yet Spencer's *First Principles*, 4th ed. (orig. ed.,1864; rev. ed., New York: Appleton, 1896), pp. 317-407 *et passim*, repeatedly stresses that the "law" of evolution is both differentiation and integration (or, more precisely, that "Evolution is definable as a change from an incoherent homogeneity to a coherent heterogeneity," p. 371); that organic evolution entails both the increased specialization of individual organs, as well as the working together, or integration, of these organs in complex whole organisms; and that the integrative pattern in the evolution of tools, moreover, is obvious in "the progress from rude, small, and simple tools to perfect, complex, and large machines.... We see that in each of our machines several of the primitive machines are united into one. A modern apparatus for spinning or weaving, for making stockings or lace, contains not simply a lever, an inclined plane, a screw, a wheel-and-axle, joined together; but several of each integrated into one whole," (pp. 334-335). (It should be noted that Spencer's assertion that universal principles of evolution are "exemplified with equal clearness in the evolution of all products of human thought and action, whether concrete or abstract, real or ideal," p. 357, precedes Popper's work in this area, and the present inquiry, by more than a century.) Indeed, it is perhaps possible to explain the differentiation/integration of tool evolution in a framework comparable to the one presently proposed for the evolution of media: (a) in the pre-technological or "pretool" situation, all human action upon the environment stems from the human body (hands, teeth, etc.) , an integrated but limited mechanism; (b) primitive tools increase human power, but only by extending individual human functions in isolation, i.e., differentiation (e.g., human hands can both sever and pound; knives can only sever, and mallets can only pound); (c) advanced tools seem to not only extend human p ower, but reintegrate previously differentiated functions into unified systems (e.g., modern food processors sever, pound, grate, etc.). See also p. 60 above, and text and notes immediately following the present note.

16. See "Teilhard's Noosphere" later in this chapter, and "Co-evolution and Convergence," with special attention to note 8, in Chapter 6 below;

see also "Talkies" and "Television" in Chapter 5 below. Note also that transportation technologies seem to be evolving towards more integrated forms, e.g., the automobile combines the flexibility of the horse with the power of the train, and air-travel can accomplish just about everything that land-travel and sea-travel can do, plus a good deal more. For more on the similar development of communication and transportation technologies, and explanation of the probable source and outcome of this convergence, see "The Reunion of Talking and Walking" in Chapter 7 below.

17. For perhaps much the same reason (and as might be expected on the basis of n. 12 above in this chapter), the evolution of language also seems to be a largely integrative process. Alexander Marshack, for example, in The Roots of Civilization (New York: McGraw-Hill, 1972), p. 117, suggests that "once voice and brain had evolved to the point where the hominid could utter syllabic words, these words would not, in the early stage, have been used as [independently] *defined* symbols, abstracted in meaning.... On the contrary, it would seem that the words would, in large measure, have been used *as part* of a communication of meaning that could only be understood within a process or relation . . . " (Italics in original.) Earliest human language, in other words, was apparently keyed to specific, immediate physical surroundings and psychological states (Marshack, p. 117, gives the example of "a cry or specialized word of warning at the presence of a carnivore"; such a cry might also communicate a feeling of fear), and only subsequently evolved the ability to generalize, through abstraction, across diverse objects and situations (e.g., the word "lion" refers to the general species of lion, or any lion, regardless of whether or not one is physically present). (Roger Brown reports the findings of numerous investigators that children usually use "concrete" words such as "milk" and "water" before more "abstract," general terms like "liquid"— though he cautions that the reverse, i.e., use of "fish" before "perch" and "bass," is also occasionally the case, and that vocabulary acquisition in children is in any event largely a function of the parent's deliberate use of "simple" words when speaking to young children. See Brown's "How Shall a Thing Be Called?" in *Readings in Child Development and Personality*, eds. Paul H. Mussen, John J. Conger, and Jerome Kagan [New York: Harper &

Row, 1965], pp. 267-276.) Thus, the evolution of media from differentiated to more integrated forms seems to parallel the much earlier evolution of language from specific to generalized usage, but with one highly significant distinction: whereas media achieve generalization through combinations of increasing speed, permanency (extension across space and time) and literal transcription of reality, language generalizes by abstraction, or by removal from literal aspects of reality; television, for example, can communicate the concept of "people" by flashing hundreds of literal faces upon the screen in 30 seconds, whereas the word "people" communicates about the general group of humans only by ignoring their hair color, facial expressions, etc. In effect, then, abstraction may be viewed as a primitive "technology" for generalization (though it of course may have other uses and consequences, such as the generation and communication of "ideas" which never existed in the real world in the first place); and its recent supplantation, in part, by more replicative technologies may be thus viewed as a switch to technologies which generalize with less sacrifice of reality. For more on the relation of abstract and replicative technologies, see "The Persistence of Abstraction" in Chapter 6 below; for a description of both language and technology as "a kind of bridging process, a way of getting from one kind of experience to another," see Marshall McLuhan, "Laws of the Media," with an Introduction by Paul Levinson, *et cetera* 34 (June 1977): 175-176.

Popper's thinking about the evolution of human products, and, more specifically, of human knowledge, thus bears much literal applicability to the evolution of media, perhaps more than Popper himself has yet considered. Indeed, Medawar has already credited Popper's World Three model with providing philosophic support for Medawar's view of technologies as evolving like biological organs and organisms.[18] Moreover, in emphasizing the adaptive context of such organ-like evolution — that is, in suggesting that not only do theories and technologies evolve in a process comparable to the evolution of organs, but, like organs, theories and technologies survive, and are modified, based on how well they serve their master organisms —

Popper, like Hall and Butler before him, makes it not at all surprising to find that media are evolving towards greater consonance with human communication environments.

18. Medawar, pp. 25-28.

That the relationship between a man and his camera is not all that different than the relationship between a man and his eyes in many ways captures the view of the human/media relationship that pervades the present work: while a man is no doubt dependent upon his eyes as presenters and processors of information (as we are upon media), it seems foolish to suggest, as Ellul and other extreme media determinists have about media, that our eyes "control" or even dominate us. Popper further clarifies this relationship between humans and their products when he seeks to distinguish between his World Three framework and the role of ideas in Hegel's dialectic, suggesting that while both view ideas as autonomous (they exist independently of their original creators, can be utilized by other humans long after their original creators are dead, etc.) and capable of influencing human behavior, in Hegel's model the autonomy and influence of knowledge is exalted to the point of overwhelming its human transmitters. Popper writes:

> Thus what I have called the autonomy of the third world [World Three], and its feed-back effect, becomes with Hegel omnipotent.... As against this I assert that the individual creative element, the relation of give-and-take between a man and his work, is of greatest importance. In Hegel this degenerates into the doctrine that the great man is something like a medium in which the Spirit of the Epoch expresses itself. [19]

It should be noted that Popper here is attacking the

motivating force of Hegel's dialectic and not its validity as a descriptive pattern (which, as suggested in the discussion of Innis early in this dissertation, seems applicable to the evolution of media), and that Popper's assertion of the "give-and-take" relationship of "a man and his work" is useful regardless of whether or not it is a warranted criticism of Hegel's perspective.[20] This assertion in effect insists upon the fine point, often missed by media theorists such as Ellul, Mumford, and even McLuhan, that human creations can influence their creators without hypnotizing or enslaving them. Such is precisely the interplay between humans and media observed by anthropotropic theory, which denies not that our media extensions influence our lives, but that such influence paralyzes or disrupts our ability to evolve these media extensions to more human functions. The anthropotropic emphasis on the human/media two-way street, with humans having the ultimate right of way, is crucial because it explains rather than ignores much of the findings of media determinism (such as McLuhan's observations on the impact of phonetic writing upon civilization), while accounting for the vast changes that media themselves have undergone. Humans may indeed be the medium through which media express themselves, but the expression itself nonetheless grows increasingly human.

19. *Objective Knowledge*, pp. 125-126.

20. Hegel's model postulates an "Absolute Spirit" (Truth) which is both a composite product and a directing goal of human expression. Such a perspective is perhaps only a bit more deterministic than Popper's view that humans create ideas and other artifacts in response to the truth of situations: in both models, Truth itself changes due to human expression; human thinking, feeling, etc., itself contains elements of the Absolute Spirit or Truth; and individuals often in effect seem mere conduits in this great interchange of forces (though here it is Popper who argues strongly that individuals have initiative in the generation of their ideas, and

Hegel who doesn't). Moreover, as already implied at the end of Chapter 3 above, there seems no necessary contradiction in proposing a theory of media evolution that is both Darwinian (Popperian) and Hegelian — that is, in suggesting that the selective mechanism of media evolution is Darwinian, while the results of this evolution fall into a pattern of thesis, antithesis, and synthesis. For more on Hegel and the dialectic, see J. N. Findlay, *Hegel: A Re-Examination* (London: George Allen & Unwin, 1958), pp. 58-82.

Perhaps the main point of difference between Popper's system and the anthropotropic model is that the evolution of ideas in Popper's system, unlike the development of media in the anthropotropic model, is essentially unpredictable. That is, Popper observes and describes the process by which knowledge grows, without ever being able to know its outcome. Such unpredictability arises from two fundamental and interrelated components of the Popper model, the first more profound than the second:

(1) Knowledge evolves, by process of error elimination, towards the "truth" of situations, and, by implication, to more general and perhaps "universal" truths. But in the falsificationist system it is only falsity, not truth, that is ever logically knowable (this follows from the inadmissibility of induction: seeing a thousand, even a million, white swans does not guarantee the general truth that all swans are white; but seeing one black swan in this situation provides certain knowledge that all swans are not white). Errors or falsities can thus be identified with certainty and eliminated with confidence from the body of objective knowledge; and purged in this way, the body of objective knowledge is left with a higher "residue" of truth content, and thus progresses towards closer approximation of the truth. But the truth itself — and hence the ultimate disposition of the evolution of knowledge towards the truth — still remains improvable, and hence unknowable (but not necessarily unattainable:

knowledge may, through elimination of all possible errors, perchance obtain a perfect correspondence to the truth; but the shortcomings of induction will prevent us from ever confirming such perfection). Truth is thus pursuable but never apprehendable — Popper entitles one of his lectures "Epistomology Without A Knowing Subject"[21] — and the evolution of knowledge towards the unapprehendable must perforce be indeterminable.

21. Objective Knowledge, p. 106.

(2) Even were it not for this most inherent of blockages, the outcome of evolution in Popper's system would be difficult to determine due to what Popper has termed the "feed-back effect," or the capacity of World Three knowledge to alter its World Two and One creators. Knowledge, once again, evolves towards the "truths" of World One, in the generator of World Two; but since perception, thinking, and feeling (World Two) can occur only through the filter of World Three knowledge, and since the truths of World One are often transformed by World Three knowledge (the discovery of antibiotics alters the nature of pneumonia), truth becomes a moving target, aimed at by a moving marksman. Every bit of knowledge, in other words, potentially changes the conception and projection of all subsequent knowledge; and in a system of all variables and no constants, prediction of outcomes or even specific directions must be a hazardous, if not impossible, undertaking.

It is only the second of these two limitations that seems applicable to anthropotropic theory, and even this in only a weak way. The lessons of media determinism demonstrate the undeniable impact that technologies have upon their human creators; but the evidence which led to the present inquiry — the advent of color TV, holography, and so forth — suggests that whatever the "feed-back" effect of media upon humans, it hasn't been strong enough to derail the human impulse to shape media in progressively closer approximations of the human situation. Thus, while the anthropotropic model must be alert to possible deflections of media evolution due to media "feed-back" or determinism (see the end of Chapter 2 above, also below discussions on " Persistence of Abstraction" in Chapter 6), such feed-back need not cripple the prediction of media evolution along anthropotropic lines.

As for the inherent unknowability of truth which serves as the first and most profound obstacle to prediction in Popper's system, the truth to which media evolve in the anthropotropic model is the "truth" of pre-technological communication — and as such, is easily and eminently knowable (through observation of history or current non-technological communication, e.g., observing two people talking to each other in person). Thus prediction of the outcome of technological evolution becomes not only possible, but the basis of the anthropotropic model.

In effect, then, the main distinction between the Popperian and anthropotropic theories of evolution is that Popper's remains, in the end, much closer to its Darwinian "open-ended" analogue; while anthropotropic theory operates on a more Teilhardian, or teleological premise (see the end of Chapter 3 above). Or, in the terminology of structuralism, the "deep structures" which Popper's objective knowledge attempts to approximate are so obscure as to be unknowable, while the deep structures that anthropotropic media attempt to recapture are eminently observable once you know what to look for. (Cf. Chomsky's view, note 3 above in this chapter, that certain "deep structures" of grammar may be inherently unknowable.)

Taken together, the structuralist perspective and Popper's conception of World Three offer a rather complete and satisfying metaphor for the world of media: the first helping to define the human-technological relationship, the second helping to describe, in most ways, its evolution. The present chapter can thus now turn from the analogues which must be translated for anthropotropic usage, to a model of communication and technology which, while not primarily concerned with either technological extension or evolution, nonetheless offers several direct and important lines of support for anthropotropic theory.

NORBERT WIENER AND CYBERNETICS: THE "WHY" OF ANTHROPOTROPIC EVOLUTION

Cybernetics, the discipline named and founded by Norbert Wiener in the late 1940s to study and improve the operation of "computing machines," has grown to occupy perhaps a unique position among theories of communication. As evidenced in the previous discussions, most models of communication — including the anthropotropic one — have borrowed heavily in conception and terminology from models in the biological, social, and physical sciences. In the case of cybernetics, however, quite the reverse seems to have occurred: concepts such as "feedback," "input," and "entropy," originating in the study of machine communication but used in cybernetics to describe living communication as well, have been quickly snapped up by biological, social, and even philosophical theorists (see, for example, Popper's "feed-back" effect in the preceding section), who appreciate the great strength of the parallel that cybernetics has drawn between machine communication and communication in living organisms. At the same time, the cybernetically-derived study of "information" and its properties have had much bearing on such primary physical problems as the properties of energy, with the second law of thermodynamics, for example, being equally applicable to information as to "physical" energy.[22] Indeed, to the extent that information can be considered a form of energy, it is possible that solutions to some of the most vexing problems in the physical universe may come from models initiated by information theory. But such is a story for another time.

22. See, for example, "Entropy and the Second Law of Thermodynamics" in Kenneth W. Ford, *Basic Physics* (Waltham, MA: Blaisdell, 1968), pp. 430-457. "Every fundamental law of nature is characterized by remarkable generality," Ford writes, "yet the second law of thermodynamics is unique among them in that it finds direct application in a rich variety of settings, physical, biological, and human. In mentioning trays of coins, molecules of gas, and disorder in the house, we have touched upon only three of a myriad of applications. Entropy and the second law have contributed to a discussion of the behavior of organisms, the flow of events in societies and economies, communication and information, and the history of the universe. In much of the physics and chemistry of macroscopic systems , the second law has found a use . . . . It is a startling and beautiful thought that an idea as simple as the natural trend from order to disorder [i.e., entropy or the second law] should have such breadth of impact and power of application," p. 440. It is worth noting here, incidentally, that Wiener views human and living systems— and, by implication, their extensions via technology— as antientropic "enzymes," or forces that work against the trend of order to disorder. See Norbert Wiener, *Cybernetics*, 2nd ed. (New York: M.I.T. Press and John Wiley, 1961), pp. 58-59; see also discussion below in the present study.

For the present inquiry, the most pertinent aspect of cybernetics is the fundamental sameness of communication it observes in technologies and living organisms. Beginning with the recognition that computers function very much like human brains — "that the logic of the machine," in Wiener's words, "resembles human logic"[23] cybernetics expands to an awareness that all communication, whether the tropisms of plants, the nervous systems of animals and humans, or the electrical circuits of machines, share a basic identity, a profound coincidence in process and function, if not precisely in physical composition and form. Thus, Wiener explains that scientists in a variety of disciplines soon "became aware of the essential unity of the set of problems centering about communication, control, and statistical mechanics, whether in the machine or in living tissue."[24]

23. *Cybernetics*, p. 126

24. Ibid., p. 11.

But this coincidence of all communication processes is far from "coincidence," in the colloquial usage of the term. For cybernetic theorists see the unity of communication as an inevitable consequence of the properties of information, which they increasingly study: "the fundamental notion," Wiener points out, is "the message, whether this should be transmitted by electrical, mechanical, or nervous means."[25] In other words, since both chemical nervous systems and electrical circuits are engaged in the processing of information — information which is subject, regardless of specific "content," to the same properties of feedback, interference, and so forth — it stands to reason that in attempting to process this information as efficiently as possible, in contending with the same functional conditions, biology and technology would attain analogous solutions.

25. Ibid, p. 8

Now such a perspective, which might be termed an "information determinism," might seem in conflict with the anthropotropic model. However, unlike media determinism, the informational focus of cybernetics is not only entirely consistent with anthropotropics, but seems to suggest a fundamental "why" of anthropotropic evolution — that is, why humans increasingly retrieve pre-technological patterns in their communication machines — that goes beyond the longing for lost wombs described by Freud (see Chapter 2 above). From the informational view, the goal of biological evolution would be to achieve the most efficient transmission of information possible. Organisms that communicate inefficiently die. Thus the end result of eons of organic evolution — millenia of genetic jockeying for better communication, entailing the jettisoning of poor processes and the preservation of effective ones — would be the most efficient communication processes possible: the "natural" communication environment, or the pre-technological communication world. And since technologies created by humans strive, after all, for the same efficiency, what could be more natural than that these technologies would come to emulate, as a matter of course, the same supremely efficient patterns of pre-technological communication hammered out in the organic foundry over millions of years? The anthropotropic urge is thus more than mere narcissism: it is the inherent human recognition and response to the unity of information and its problems of transmission in the universe (although psychological pleasure is no doubt the immediate motive for choice of informationally appropriate technologies in much the same way as pleasure motivates the biologically appropriate act of sexual intercourse in humans). Humans are revealed, once again, as more than the matter of the cosmos examining itself. We may be more properly considered, in Teilhardian terms, as active if unconscious agents of the cosmos: recreating and spreading in a few short years, in our technologies, the collective and slowly-won communicative lessons of life and matter.

Viewed in such a context, anthropotropic theory becomes a special case of cybernetics — an "anthropology" of cybernetics, as it were, which details the steps in which humans have come to mold their communication machines in increasing consonance to the properties of information, and thereby in increasing similarity to the communication modes of the natural world and biology. Put more succinctly, anthropotropic theory seeks to explain how technologies have grown (and are growing) similar to pre-technological communication — a similarity which cybernetics would of course be already very much aware of. The general thrust of anthropotropic theory is thus already anticipated by cybernetics, making cybernetics the more fundamental and encompassing of the two frameworks.

The present theory, then, in effect performs a detailing service for cybernetics; but cybernetics performs a far more crucial service for the present theory: in suggesting a primary equation between efficiency and biology (pre-technology) in communication, cybernetics helps explain the at first seemingly paradoxical observation that the most primitive, i.e., least efficient, technologies are also the most unlike, and worst transgressors of, nature (these technologies have been described here as "Stage B"); while the more advanced, i.e., more efficient, technologies operate increasingly like natural communication ("Stage C" technology).[26] (Such an equation also undermines the Ellulian argument that, in making "efficiency" the supreme ethic of modern society, technology therein destroys the human or natural elements of existence.) Cybernetics thus offers much more than a metaphor for the present theory, and provides much more than a particular piece of the anthropotropic puzzle: it provides in its most general sense a central piece of the anthropotropic puzzle, or perhaps the backboard upon which the model is pieced together.

26. The equation between efficient and pre-technological communication has been empirically confirmed with findings reported by the

Communications Studies Group in England — which indicate "that even with a relatively simple task, there may be differences in communications efficiency between face-to-face and telephone . . . with no vision there were longer pauses and more interruptions . . . with no vision there were more questions, and requests to repeat. . . . " Cited by Felipe Korzenny, "A Theory of Electronic Propinquity," *Communications Research* 5 (January 1978): pp, 3-23. It follows, then, that a more advanced technology such as video-phone would be both more natural and more efficient than the telephone.

BRAINS, COMPUTERS, AND MINDS

Cybernetics also provides a more specific assistance to anthropotropic theory in stressing the growing similarities of the technological computer to the human brain. Since the computer may be considered the "master" medium — "master" in that both broadcast and interactional media such as telephone and facsimile mail increasingly rely upon computers, and government officials expect this trend to continue[27] — a convergence of the computer towards the human brain, which is the "master" of the human nervous and sensory systems, is especially noteworthy. (Indeed, the observation can be made that while all other media extend human mentality indirectly by extending our senses, the computer alone permits a direct extension of the human thought process.) Carl Sagan briefly describes the extent of this computer/human brain equivalence in *The Dragons of Eden*:

> The left hemisphere [of the human brain] processes information sequentially; the right hemisphere simultaneously, accessing several inputs at once. The left hemisphere works in series; the right in parallel. The left hemisphere is something like a digital computer; the right like an analog computer.[28]

In other words, computers have been developed to the degree that they can now approximate (or mimic) both the logical and "intuitional" (analogical) functions of the human mind. (Note that this observation stands even if we reject the contention that logic is "located" in the left hemisphere, and intuition in the fight hemisphere of the brain. Note also that computers are here approximating human mentality or the performance of the human brain, and not necessarily the brain's physiological structure— thus providing some interesting confirmation of the existence and impact of "mentality" as a force distinct from the physical brain. See also the end of Chapter 1 above.)

27. See Richard Taylor, "Towards Communications Policies for the 21st Century," *Communications Tomorrow*, September 1977, pp. 1, 3.

28. (New York; Random House, 1977), p. 169.

Of the two computer "reasoning" modes — digital and analogical — the digital seems the more likely to enhance the computer's position as the "master" medium since, as Richard Taylor points out in a recent World Future Society newsletter,

> All types of information — telephone conversations, television pictures, books, virtually anything — can be coded and transmitted in the form of discrete binary digits or "yes-no" signals . . . . such systems have technical advantages over present systems using 'analog' techniques . . .
>
> 'Digital' has always been the 'natural language' of computers, and it appears that in the future, digital bits will provide the common form for almost all information storage, transmission, switching, processing and display.[29]

Such a development would not be surprising, since it is certainly the "logical" part of human mentality that coordinates modern society, and the digital computer extends that logical mode. Of more unexpected significance, perhaps, is the disclosure that "digital" has always been the "natural language" of computers: providing a striking parallel to Levi-Strauss' assertion that the natural language or process of the human mind consists of binary opposites. Indeed, the above description of computer binary digits as "yes-no" signals could have been lifted from any textbook on structural anthropology, so close is the operation of computer binary digits to human bi-polar opposites. But Levi-Strauss' inquiry into the inherent structures and operation of human mentality has been relatively innocent of computers or cybernetics (Levi-Strauss characteristically studies myths of primitive, often pre-literate, societies),[30] thus the almost literal similarity of the most important and fastest growing type of computer operations to independently discovered principles of human mentality provides strong support for the anthropotropic perspective. The digital computer in this way makes structuralism a further ally of the present theory.

29. Taylor, p. 1.

30. See Edmund Leach, Claude Levi-Strauss, for a discussion of Levi-Strauss' techniques and subject matter. However, see Robert Blechman, "Myth as Advertising: A Structural Analysis of Selected American Television Advertisements Using a Methodology Based on the Theories of Claude Levi-Strauss" (Ph.D. dissertation, New York University, 1978), for an application of Levi-Straussian procedures to the content of a modern technology.

THE MEDIUM THAT BECAME THE WORLD: TEILHARD'S NOOSPHERE

The specific symmetry of computer and human thought processes recalls an earlier, more metaphorical parallel that has been drawn between technologies and the human brain, in which the interrelated total of all communication technologies functioning together are said to resemble a unified central nervous system, or a single, "super," human brain. Indeed, as Lewis Mumford has pointed out, novelists and philosophers were describing the telegraph and electronic media as transforming the whole earth into a "living brain" well over a hundred years before computers and cybernetics. Mumford, not surprisingly viewing such descriptions as unheeded warnings, relates the musings of both Nathaniel Hawthorne and Ralph Waldo Emerson on this matter. First Hawthorne, through his character Clifford in *The House of the Seven Gables*, rhetorically wonders,

> 'Is it a fact that . . . by means of electricity the world of matter has become a great nerve, vibrating thousands of miles in a breathless point of time? Rather, the round globe is a

> vast head, a brain, instinct with intelligence! Or, shall we say, it is itself a thought, nothing but thought, and no longer the substance which we deemed it.'[31]

And then Emerson, nearly anticipating the very terminology of the present study, concludes that "our civilization and these ideas are reducing the earth to a brain. See how by telegraph and steam the Earth is anthropolized."[32]

31. As cited by Mumford, *The Pentagon of Power*, p. 314. Notice, once again, how Hawthorne here calls attention to the technological approximation of human thought, as well as the physiological structure of the human brain.

32. Cited by Mumford, *The City in History* (New York: Harcourt Brace Jovanovich, 1961), p. 567.

But as Mumford recognizes, nowhere is the metaphor of earth-as-brain-through-technology more comprehensively and eloquently presented than in the work of Pierre Teilhard de Chardin, who, writing much closer to our own time, was in a much better position to observe it.[33] Teilhard, whose view of the universe as ascending through levels of greater complexity and consciousness to divinity has already been mentioned as a fitting model in many ways for the "ascension" of media to human modes described by the present theory (see end of Chapter 3 above) considers the shift from the individualized and separated human mentalities of earlier times to the more collective or globalized consciousness of the present and future to be a key step in the human journey towards divinity. And most instrumental in attaining such a collective mentality or "Noosphere" are communication technologies, the role of which Teilhard describes as follows:

> . . . how can we fail to see the machine as playing a constructive part in the creation

of a truly collective consciousness? It is not merely the machine which liberates, relieving both individual and collective thought of the trammels which hinder its progress, but also of the machine which creates, helping to assemble, and to concentrate in the form of an ever more deeply penetrating organism, all the reflective elements upon earth.

I am thinking, of course, in the first place of the extraordinary network of radio and television communications which, perhaps anticipating the direct inter- communication of brains through the mysterious power of telepathy, already link us all in a sort of 'etherised' universal consciousness.

But I am also thinking of the insidious [i.e., almost invisible] growth of those astonishing electronic computers which, pulsating with signals at the rate of hundreds of thousands a second, not only relieve our brains of tedious and exhausting work but, because they enhance the essential (and too little noted) factor of 'speed of thought', are also paving the way for a revolution in the sphere of research . . . .

There is a school of philosophy which smiles disdainfully at these and kindred forms of progress . . . . One is tempted to call them blind, since they fail to perceive that all these material instruments, ineluctably linked in their birth and development, are finally nothing less than the manifestation: of a particular kind of super-Brain . . . from which every individual within the collective process benefits . . . .[34]

33. See *The Pentagon of Power*, p. 314-320, for Mumford's views on Teilhard.

34. Pierre Teilhard de Chardin, *The Future of Man*, trans. Norman Denny (original ed., 1959; trans. ed., New York: Harper & Row, 1964), pp. 167-168.

Teilhard's view of technology (as distinct from his model of evolution) may be said to bear the following interesting relationship to anthropotropic theory: although anthropotropic theory can stand or fall independently of whether Teilhard's views are correct or not, there seems no reason to conclude that Teilhard's views are not correct. As indicated in the opening chapter, the present theory seeks to describe, explain, and perhaps predict the course of media evolution, without necessarily rendering a judgement as to whether this course is beneficial to our species or not; yet as will be discussed in the concluding chapter, there seem few convincing reasons why the anthropotropic evolution of media need not be greeted with at least part of the optimism that Teilhard holds for technology.

Much the same may be said about the view of Teilhard and others of the Earth as an emerging "superbrain" of technology: anthropotropic trends are at present best observed in the one-to-one correspondence of specific media to specific modes of human communication, e.g., the growing similarity of audio-visual media presentations to real-world communication of sight and sound; demonstration of a "collective" correspondence of technology to the human nervous system, in addition to the specific correspondences already observed, would thus be superfluous to the basic establishing of anthropotropic theory. However, such a theoretical stance by no means rules out the possibility (or likelihood) that media may in fact evolve in collective or interrelated networks towards approximation of human communication patterns. Several examples of apparently "cooperative" media evolution have already been mentioned above in connection with Harold Innis (see the end of Chapter 3 above) — such as the extension of the voice across time by the phonograph to complement the extension of the voice across space by telephone, attaining through two media a time/space acoustic balance that replicated the pre-technological human time/space balance more fully than either medium could on its own (see Chapter 6, below, "Coevolution and Convergence," for a further discussion of how media may in concert achieve a greater correspondence to a prertechnological communication mode than they would individually). Moreover, the advent of telecommunications satellites— barely even predicted at the time of Teilhard's writings[35] — certainly seems tangible evidence of the type of unified nervous system girdling the earth described by Teilhard.

35. Arthur C. Clarke was apparently the first to publicly envision the use of satellites for telecommunications, in a letter to the British magazine *Wireless World* published in 1945; for more on Clarke's prediction, and a brief history of telecommunications satellites, see Brenda Maddox, *Beyond Babel* (New York: Simon & Schuster, 1972), pp. 66-141. According to Birx, pp. 16-17, the bulk of Teilhard's philosophic work was done from 1937 to 1955, though owing to Vatican opposition was not published until after his death in 1955.

Thus, while the anthropotropic theory need not demonstrate "collective" media evolution to establish itself as a valid theory, it need not be unaware of collective media evolution either. In effect, this question of separate vs. collective media evolution raised by Teilhard and others may be summarized by three propositions of increasing force but decreasing likelihood (and demonstrability): (1)

individual media evolving towards greater approximation of individual or specific pre- technological modes, as in the camera functioning like the eye; (2) media cooperating to achieve a greater approximation of pre-technological communication than they do individually, as in the above example of the telephone and phonograph; and (3) all communications media interrelating to form a single, unified, magnified human brain, sensory, and nervous system, that encompasses the whole earth and (through space exploration) beyond. (Note that this system is not necessarily incompatible with the more differentiated view of proposition #1 above, that is, television cameras and microphones could be the "eyes and ears," computers the "brain," and so forth of a single, collective, super-system.) [36] It is the task of the present work to propound a theory of media evolution based on the first proposition; the second proposition appears to be an important and perhaps overriding principle or strategy of that evolution; the third remains an intriguing but as yet only incompletely emerging possibility.

36. Computers, for example, coordinated and analyzed the sensory probes in the Viking missions to Mars in 1976, which not only "saw" but "tasted" and "sniffed" the Martian environment. See Walter Sullivan, "Tests to Seek Life on Mars Begin," *New York Times* , 25 July 1976, pp. 1, 28; and Sullivan, "How to Search for Life on Mars," *New York Times*, 1 August 1976, sec. 4, p. 8. Note also the speech delivered by U.S. Vice President Walter Mondale to the United Nations on May 24, 1978 (reported by Kathleen Teltsch, "Mondale, at U.N., Offers Nations Equipment to Help Avert Conflicts," *New York Times*, 25 May 1978, pp. Al, A 1 6) , in which Mondale refers to new radar and sonar tracking systems as the "eyes and ears of peace."

Computers and unified-technologies-as-brains are surely on the cutting edge of media evolution: computers are perhaps the most recent major innovation in communication technology (with the possible exception of holography) ; and the unified brain view, as explained above, is a good deal away from being thoroughly realized. Exploration of both these emergent systems and the theories that seek to explain them, briefly attempted above, has provided some interesting leads and support for the anthropotropic thesis. This chapter will now conclude by briefly turning to the other end of the evolutionary tree — to the past of media evolution rather than its future — to examine some of what has been written about the pioneering medium of film. Motion photography is of course pre-dated by many other technological media (still photography, telephone, etc.); yet, it was perhaps the first technological medium to achieve widespread artistic recognition and generate an indigenous popular culture, and as such has long been the subject of an area of study and contemplation known as "film theory."[37]

37. Vachel Lindsay's *The Art of the Moving Picture*, published in 1915, and Hugo Munsterberg's *The Photoplay: A Psychological Study*, published in 1916, are generally considered among the first serious theoretical analyses of film. See James Monaco, *How to Read a Film* (New York: Oxford Univ. Press, 1977), pp. 298-301.

SIEGFRIED KRACAUER: FILM AS A HAPPY ENDING

The "school" of film theory of most interest to the present inquiry is, unsurprisingly, known as film "realism." Siegfried Kracauer provides an overview of this perspective in the epilogue of his *Theory of Film: The Redemption of Physical Reality*, where he postulates a relationship between reality, science, technology, and film, which, it will be observed, seems to lift a page almost literally out of Jacques Ellul, but with one highly significant twist — Kracauer's parable has a happy ending:

> We not only live among the 'ruins of ancient beliefs' but live among them with at best a shadowy awareness of things in their fullness. This can be blamed on the enormous impact of science....
>
> Most sciences do not deal with the objects of ordinary experience but abstract from them certain elements which they then process in various ways.
>
> Thus the objects are stripped of the qualities which give them 'all their poignancy and preciousness' (Dewey). The natural sciences go furthest in this direction. They concentrate

on measurable elements or units, . . . and they isolate them in an effort to discover their regular behavior patterns and relationships . . . . The social sciences . . . aim at achieving the status of exact sciences and . . . on their part aspire to further mathematization of the traces of reality. . . .

One of the main channels of this influence is of course technology. . . .

Evidently we can limit our all but compulsive indulgence in abstraction only if we restore to the objects the qualities which, as Dewey says, give them their 'poignancy and preciousness.' The remedy for the kind of abstractness which befalls minds under the impact of science is experience — the experience of things in their concreteness. Whitehead was the first to see our situation in this light . . . .'When you understand all about the sun and all about the atmosphere and all about the rotation, of the earth, you may still miss the radiance of the sunset. There is no substitute for the direct perception of the concrete achievement of a thing in its actuality' . . . .

And how can this demand be met? . . .

Film renders visible what we did not, and perhaps even could not, see before its advent. It effectively assists us in discovering the material world with its psychophysical correspondences. We literally redeem this world from its dormant state, its state of virtual nonexistence, by endeavoring to experience it through the camera. . . . The cinema can be defined as

> a medium particularly equipped to promote the redemption of physical reality. Its imagery permits us, for the first time, to take away with us the objects and occurrences that comprise the flow of material life.[38]

Thus, to Ellul's vision of a world abstracted to alienation through technology, Kracauer answers: "film!" (Of course, Ellul would surely find the very "taking away with us" via film of the "objects and occurrences that comprise the flow of life" to be merely another, no doubt even more insidious, disruption of that flow of life.) Kracauer's views, however, would not be alien to McLuhan, who, as suggested above (end of Chapter 2), sees the convergence of linearity through phonetic writing, print, science, and the industrial technologies as profoundly distortive and disruptive of human sensibilities — and finds the remedy in new electronic technologies that tend to restore sensory balances (film performs like an electronic medium in that it provides a literal, rather than abstract, presentation of sensory material).

38. (New York: Oxford Univ. Press, 1960), pp. 291, 292, 296, 300.

Nor would Kracauer's model be unfamiliar to anthropotropic theory: indeed, the destructive "abstractions" of science and technology that Kracauer rails against are easily recognizable as the price paid for what has here been termed "Stage B" or primitive, trade-off technologies; and the "redemption of physical reality" through to film, a splendid example of the anthropotropic or "Stage C" retrieval of that previously sacrificed real world.

One serious difference between Kracauer's and the present perspective is that Kracauer seems to reserve the role of rescuer solely to film (this includes film's antecedents such as still photography, and audio-visual media such as television which "share certain essential characteristics" with film and help make films more available to the public). [39] Thus, Kracauer makes a special point of criticizing "certain media of communications [where] . . . the transmitting apparatus overwhelms the contents transmitted,"[40] thereby failing to retrieve reality and merely adding to the technological malaise. Radio is judged particularly culpable in this regard, merely catering to people's loneliness with "noise," rather than genuinely alleviating the loneliness with reality. Such a position, moreover, may not be totally unjustified: as suggested at the outset of the present work, film and its adjuncts are certainly the earliest, and most vividly developed forms of anthropotropic media (Kracauer is by and large correct in suggesting that television doesn't so much add to the realistic content of film as give it a fuller distribution; see the end of Chapter 5 below). Kracauer can be faulted, however, for failing to see film as a vanguard of a more universal media evolution — ignoring the significance of the computer as just discussed above, to cite one example, and misreading radio in apparently not noticing that it redeems a fuller sensory reality than prior media such as print.

39. Ibid., pp. 166-167.

40. Ibid., p. 293.

ANDRÉ BAZIN: THE ANTHROPOTROPIC INTENTION

Such oversights have by no means been shared by all film "realists"; thus André Bazin, who along with Kracauer is considered the main founder and exponent of the "realism" school,[41] sees the drive to replicate reality as underlying all reproductive (and presumably communicative) technologies. The single, guiding desire, Bazin explains,

> inspiring the invention of cinema, is the accomplishment of that which dominated in a more or less vague fashion all the techniques of mechanical reproduction of reality in the nineteenth century, from photography to the phonograph, namely an integral realism, a recreation of the world in its own image. . . . [42]

41. Monaco, p. 306, points out that "Kracauer has the reputation of being the foremost theorist of film realism," adding that although André Bazin "offered a much richer investigation of the phenomenon during the fifteen years preceding Kracauer's book [*Theory of Film*] , his work was never so clearly codified as Kracauer's and therefore hasn't until recently had the direct impact of his successor's." Gerald Mast and Marshall Cohen agree in *Film Theory and Criticism* (New York: Oxford Univ. Press, 1974), pp. 2-3, that Bazin is a "second generation" film realist, although his work

predates the "first generation" Kracauer.

42. André Bazin, *What is Cinema?*, with a Foreword by Jean Renoir, comp, and trans. Hugh Gray (original ed., 1958-1965; trans. e d . , Berkeley and Los Angeles: Univ. of Cal. Press, 1967), p. 21.

But Bazin's main contribution to anthropotropic theory goes far beyond merely affirming (or anticipating) its scope: for he tackles the fundamental question of why, if technologies spring from the human desire to replicate reality, so many technologies are at first such poor replicators of reality; and in addressing that question uncovers a rather obvious (once uncovered) but often neglected fact about the emergence of new technologies, namely, that

> If cinema in its cradle lacked all the [replicative] attributes of the cinema to come, it was with reluctance and because its fairy guardians were unable to provide them however much they would have liked to.[43]

In other words, technologies that fail to replicate reality do so not by design, but inadequacy of design— that is, the inability of technological know-how at the time to fulfill the goals of inventors and consumers. That the human desire to fully retrieve reality is present at the start is substantiated by Bazin with historical evidence such as

> . . . Edison with his kinetoscope made to be attached to a phonograph, or Demenay with his talking portraits, or even Nadar who shortly before producing the first photographic interview, on Chevreul, had written, 'My dream is to see the photograph register the bodily movements and facial expressions of a speaker while the phonograph is recording his speech' (February, 1887). If color had not yet appeared, it was because the first experiments

with the three color process were slower in coming. But E. Reynaud had been peiinting his little figurines for some time and the first films of Melies are colored by stencilling.[44]

43. Ibid.

44. Ibid., p. 20

Such observations actually echo an "anthropotropic" principle first noted by Sergei Eisenstein (not generally regarded as a theoretical "realist")[45] in his essay "Not Coloured, But in Colour," where he points out that "sound sprang from the inner urge present in the silent film to go beyond the limits of plastic expressiveness alone" and "as the silent film cried out for sound, so does the sound film cry out for colour." [46] This Eisenstein/Bazin principle in effect contains three important and interrelated lessons for anthropotropic theory: (1) the desire to fully retrieve reality (or conserve the pre-technological communication environment) is already present at the invention of the first technologies; (2) partial or imperfect retrieval of reality occurs only because the technology to fully retrieve reality is not practically feasible at the time (or: lack of technical know-how limits early technologies to extension without full replication — that is, technological extension of reality can at first take place only through abstracting or otherwise distorting that reality out of its original context); (3) partial retrieval of reality is hence only a transitory phase (its duration dependent upon when the next technological innovation will be made). These observations have certainly been already expressed in one way or another in the course of the present study; but their presence in the work of Eisenstein and Bazin — especially in Bazin's recounting of fuller replicative goals in the minds of partially replicative inventors — helps bring the principles into the sharper focus of tangible events and media. (See Part II below for a more concrete treatment of the evolution of media.)

45. Monaco, pp. 301-302, 309, for example, distinguishes between film "realism," in which the goal is to capture the "raw materials" of reality, and film "expressionism," in which the goal is to "modify or manipulate reality" so as to emphasize "the filmmaker's expression of the raw materials [rather than] . . . the filmed reality itself." Under these standards, Eisenstein, whose emphasis of montage "has as its aim the creation of ideas, a new reality, rather than the support of . . . the old reality of experience," would obviously classify as an expressionist rather than a realist.

46. Sergei Eisenstein, *Notes of a Film Director*, with a Note by Richard Griffith, comp, and ed. R. Yurenev, trans. X. Danko (original ed., n.d. [shortly after 1948]; new ed., New York: Dover, 1970), pp. 114, 116.

The view that film is a replicator of physical reality is not, it must be noted, shared by all — or even a majority, it seems — of film theorists. Thus, Andrew Tudor makes a representative criticism in his *Theories of Film* when he asks why, if film is indeed a replicator of reality, (a) most films are fiction, not documentaries, and (b) even in documentaries and documentary-like films such as *Citizen Kane*, there are numerous "distortion[s] by camera angle . . . destroying the visual unity of space . . ."[47] As Tudor himself seems to recognize, the first criticism is easier to answer than the second: the reality of film is obviously not its story or subject-matter but its structure.[48] Thus, fiction in film makes film no less a replicator of reality than telling a lie in a telephone conversation would render the telephone "unrealistic," or, for that matter, lying in a real-life interpersonal conversation would somehow render the real-world "unreal." Moreover, it is probably the very structural reality of film that makes film fiction so effective: for fiction to at all succeed, it must at least be believable — the audience must accept the fictional situation as a possibility — and the presentation of fiction in realistic structures (e.g., color, sight-and-sound, etc.) certainly must enhance its believability. (As I and others have argued elsewhere, all successful art appears to have some grounding in patterns of living and reality.)[49] As for Tudor's second objection to film realism — that camera angles and sequences often seem destructive of "real-life" flow — it seems possible that Tudor may be adopting a too narrow definition of "real-life": in the pre-technological world, viewer mobility allows for perception of numerous and rapid-changing angles and sequences. The movie house, however, robs the viewer of this mobility (and so, for that matter, does television — i.e., you can't walk around in back of the screen to see the back of the subject's head), so might not the multiple angles of film be a way of restoring the multiple angle possibilities of real-life (i.e., presenting the back of the subject's head to the immobile viewer)?[50] Such is only speculation, however; and Tudor's criticism in this regard, as a criticism relating to film structure rather than content, merits further attention (see Chapter 5 below).

47. (New York: Viking, 1973), pp. 111, 114.

48. Thus, Tudor, pp. 108-111, correctly relates Bazin's view that the "reality" of film lies not in its presenting material which is factually "true" or even "plausible," but in its "'straightforward photographic respect for the unity of space' "— that is, the physical structure and organization of the real world. (The Bazin quote is from *What is Cinema*, p. 46.)

49. Such may seem to contradict the definition of art at the end of Ch. 1 above as a "transcendence" of reality. Indeed numerous theorists, of which Mast and Cohen, pp. 1-2 , are representative, draw a sharp distinction between those who, like Aristotle and Shakespeare, view art as a "mirror up to nature" (*Hamlet*, act 3, sc. 2, line 24), and those who, like Langer et al. cited in Ch. 1 above, view art as a transformation of reality or nature. Such a contradiction, however, is by no means warranted: forthe very act of holding the mirror up to nature implies going beyond

nature (no intentional mirrors exist in nature), and the very act of transcending reality entails some origin in reality. Art, in other words, is more than merely non-real; it is a purposeful departure from the real which, to be effective, must have some discernible connection to that which it has departed from (for to do otherwise would no longer be to depart from or transcend, but to merely obviate or obliterate). Such was recognized as long ago as 1833 by John Abercrombie, who writes in *Inquiries Concerning the Intellectual Powers, and the Investigation of Truth* (original ed., n.p., 1833; reprint ed., Boston: Otis, Broaders, 1843), p. 129, that "in the exercise of imagination, we take the component elements of real scenes, events, or characters, and combine then anew by a process of the mind itself, so as to form compounds which have no existence in nature. A painter, by this process, depicts a landscape combining the beauties of various real landscapes, and excluding their defects. A poet or novelist, in the same manner, calls into being a fictitious character. . . . The compound in these cases is entirely fictitious or arbitrary; but it is expected that the individual elements shall be such as actually occur in nature, and that the combination shall not differ remarkably from what might really happen. When this is not attended to, as in a picture or a novel, we speak of the work being extravagant or out of nature. But, avoiding combinations which are grossly at variance with reality, the framer of such a compound may make it superior to any thing that actually occurs." Such was also, in effect, the position taken in my "Toy, Mirror, and Art: The Metamorphosis of Technological Culture," *et cetera* 34 (June 1977): 151-167; and in Ch. 1, n. 30 above. For the purposes of the present discussion, this view of art helps explain how realistic technologies can enhance the presentation of more-than-real art.

50. A similar explanation for the apparently "unnatural" rhythm of some TV sequences has been suggested by Joshua Meyrowitz, who writes that "although the 'cutting' [abrupt switching] of shots is often thought of as a distortion of perception because we do not see the space in between (as we would in a 'pan' [continuous sweep] shot), the cut actually closely resembles an individual's own scanning of a live scene. If I am watching two people talking in a live encounter, I will often look at one and then the other. I rarely attend to the space between them. In effect, I see in

'cuts,' not pans." See "Television and Interpersonal Behavior: Codes of Perception and Response," in *Inter/Media: Interpersonal Communications in a Media World*, eds. Robert Cathcart and Gary Gumpert (New York: Oxford Univ. Press, forthcoming). One might argue, then, that a primary purpose of editing in film and "switching" in television is to compensate for the flexibility that the camera removes from the human eye.

However, even if an aesthetic realism of film is rejected (i.e., if the very distortions and alterations of reality still present in film are judged the prime causes for its art), Kracauer's view of film as a literal antidote for the disease of over-abstraction, and Bazin's (and Eisenstein*'s) description of the necessary evolution of motion, sound, and color photography, still seem difficult to deny, and thus make important contributions to the building of a general anthropotropic theory. And Bazin's observation of the "anthropotropic" intentions of many inventors offers special assistance, as indicated above, in explaining the inadequacies of early technologies as undesired shortcomings rather than endemic and permanent features.

* * *

These contributions can now be added to the lessons garnered from the other theoretical perspectives considered in the present chapter, which may be summarized and enumerated as follows:

(1) Chomsky's construct of "deep structures" of grammar, which can serve as a useful metaphor for the structures or patterns of pre-technological communication, calling attention to the deterministic relationship of pre-technological structures to technologies, and the probable genetic derivation of pre-technological patterns. (2) Levi-Strauss1 view of human mentality as processing information into "bi-polar opposites," which helps account both for Innis' observation of succession of extensional

opposites in early media evolution, and the growing tendency for simultaneity of opposites both in individual media and over-all media environments. (3) Popper's view that products of the human mind (which include communications media) evolve in a Darwinian-like fashion.

Since none of the above perspectives directly deal with communication technologies, they are actually only analogues for the present theory, subject to at least the two following limitations: (4) The transformations from deep structures to surface manifestations in both Chomsky's and Levi-Strauss' models are relatively fixed, whereas it is the very essence of anthropotropic theory that the reflections of pre-technological patterns ("deep structures") in technologies ("surface structures") are evolving, to wit, becoming more accurate or complete. (5) Levi-Strauss' and Popper's systems deal primarily with "ideas," which are more easily and rapidly generated, and more unpredictable (in Popper's system), than are technologies in the anthropotropic model.

Cybernetics is directly concerned both with communication of information in general and with computers in particular, and thus is subject to none of the analogical limitations listed above. The cybernetic perspective makes at least two important contributions to anthropotropic theory: (6) The properties of information are the same whether transmitted by living tissue or technologies, thus both living organisms and technologies develop comparable modes of communication. From this principle, anthropotropic theory can deduce two important consequences: (a) living organisms (the pre-technological world) have developed the most efficient communication modes to date, since they have had billions of years to either develop such modes or die; (b) in attempting to build more efficient technological communicators, humans thus inevitably construct technologies that are more like living or pre-technological systems. (7) As a particular case in point, cybernetics emphasizes the growing functional resemblance of computers to the human brain; of special note is the observation that computers are becoming the "master switchers" of the media world, much as the human brain is the master switcher of the human sensory and nervous systems.

As an extension of the computer-as-brain comparison, Teilhard and others suggest that technological systems as an interrelated whole are forming one super-brain around the earth; such a proposition seems still somewhat unrealized at present, but calls attention to an important possibility that anthropotropic theory must consider: (8) the collective or interrelated evolution of media towards replication of pre-technological patterns.

Finally, film theory, which analyzes a particular medium and thus like cybernetics is more than a analogue, makes three contributions to the present work: (9) the view of film as a cure for over-abstraction, (10) the inevitability of motion, sound, and color photography, and (11) the anthropotropic intentions of inventors — as explained more fully above.

In concluding a consideration of these analogues and particulars, the present inquiry completes a series of three chapters which have sought out the partial and prior expressions of anthropotropic principles in a variety of theoretical perspectives. The discussions of these works here suffered from the usual inadequacies of summaries: students of specific disciplines will no doubt feel that much of the subtlety and perhaps even major implications of certain theories were unaddressed; and in some instances a theoretical approach has been presented as the work of a single theorist, when in fact the approach was the product of numerous theorists (Teilhard's model of evolution, to cite but one example, was certainly anticipated in many ways by "process" philosophers such as Samuel Alexander and Alfred North Whitehead, and "creative" evolutionist Henri Bergson. [51] However, ideas and theorists were singled out here not only as representatives of often larger bodies of thought, but because the ideas were deemed central to these bodies of thought, and the theorists were the originators or at least providers of the most fully developed formulations of these perspectives.

51. See Franklin L. Baumer, Modern European Thought (New York: Macmillan, 1977), pp. 450-453, for a brief discussion and comparison of Bergson, Alexander, Whitehead, and Teilhard.

Aspects of theories were of course also selected on the basis of their adaptability to anthropotropic theory; and yet the one observation that can be made about all the theories considered above is that not one of them offers a full expression of the simple anthropotropic view that media are evolving towards increasing replication of human environments. Thus, even McLuhan, the most profound of media theorists, notices a three-part pattern of oral, print, and electronic technologies, but fails to see this pattern as a single, continuing process of evolution; while Medawar describes the Darwinian-like evolutionary character of technologies, without providing a clear picture of the direction of that evolution. However, it is in the sifting through of these various frameworks for pieces that fit— and, at least as importantly, in the attempt to explain why certain pieces don't fit— that anthropotropic theory has thus far been developed and refined. From psychologist Freud comes the view and discussion of media as manifestations of human desires; from anthropologist Hall the view of technologies as biologically adaptive; from historian Innis the importance of time and space extensions; from linguist Chomsky the possibility of genetically-determined technology; from philosopher Popper's criticism of Hegel an elegant statement of the shortcomings of media determinism; from mathematician Wiener the groundwork for an equation between natural and efficient communication; from theologian Teilhard's model of God and humanity a model that helps explain humanity and technology; and so forth. Thus imbedded in so wide a panoply of perspectives, the half-recognized outlines of the anthropotropic theory are perhaps its strongest proof; yet thus fortified with such insights, the present inquiry now turns to document their operation by examining the actual historical evolution of technological media.

PART II

THE ANTHROPOTROPIC EMERGENCE

CHAPTER 5

ORIGINS AND ASCENT OF THE SPECIES

If technology were to be defined in terms of industrial or mass production, then the printing press might well be considered the first technological communication medium; and, indeed, many college texts on "mass media" begin their historical accounts with a discussion of "the book" or one of the other early products of the printing press.[1] But since the present inquiry is interested in "technological" rather than merely "mass" media — having already defined "technological" communication as that being aided, abetted, or made possible through the intervention of an artificial contrivance, which leads to the distinction that all mass media are technological, but not vice versa — the present inquiry must surely take note of technological communication that vastly pre-dated the printing press, such as scribbling on a piece of parchment. (Indeed, since spoken language itself is both a cultural artifact — at least in so far as vocabulary is concerned — and an obvious communicator, it too can perhaps be considered a communication "technology." Thus Popper views spoken language as a pre-eminent "product" of the human mind, a product which may make all other products, and perhaps even the concept of "self," humanly possible;[2]

and McLuhan suggests flatly that "all our artifacts, all our 'sensory and motor accessories,' are in fact, *words*." [Italics in original.][3] (See also discussion on speech as a technology in Chapter 6 , pp. 249-251 below.)

1. For example, Ray Eldon Hiebert, Donald F. Ungurait, and Thomas W. Bohn's *Mass Media* (New York: David McKay, 1974), which according to the publisher is used in hundreds of colleges and universities in the United States, begins its survey of mass media with "the book."

2. "One could say," Popper suggests in the concluding words of *The Self and Its Brain*, p. 566, "that man has created himself, by the creation of descriptive language and, 'with it, of World 3 [i.e., products of the human mind such as theories, technologies, etc.]." See Ch. 4, n. 9 above, and the concluding part of n. 20 in Ch. 6 below.

3. "Laws of the Media," p. 175. The equivalence of speech and technology, McLuhan further contends, is recognized by at least two African tribes who use the same terminology to describe the acts of speaking and weaving. McLuhan quotes from Douglas Fraser's *African Art as Philosophy*: " Among the Bambara and the Dogon, the gift of weaving is closely associated with that of speech. 'Soy,' the Dogon word for cloth, means 'It is the spoken word .' Weaving, along with speech, was a gift from the Creator to help man." It might added that in our own culture, of course, the connection between speaking and weaving expresses itself in the metaphor, "spinner of tales."

HIEROGLYPHICS AND PHONETIC WRITING: THE PRIMITIVE SYNDROME

The origins of writing are of course a lot less clear than the first usages of more modern media, yet two features of writing are nonetheless immediately apparent: (a) writing as a whole was (and is) a supreme example of primitive "Stage B" or trade-off technology; (b) the evolution within writing from hieroglyphics to the more abstract phonetic alphabet appears to contradict the anthropotropic thesis.[4] It is unfortunate to come upon an apparent contradiction at the very origins of technology, yet perhaps being situated at the origins is the very cause of the contradiction. Hieroglyphics and other varieties of non-phonetic writing, among the first communication technologies ever devised, performed poorly both as replicators of reality (even the most primitive, literal pictographs are highly stylized visual representations at best, in which sounds, etc., are of course lost) and as extenders across time and space (picture writing takes a relatively long time to learn and to execute, occupies a relatively large amount of space per information conveyed, etc.).[5]. Yet extension across time and space is precisely the raison d'etre of "Stage B" technologies — thus, hieroglyphics may be considered a "Stage B" technology that failed in its fundamental task, that is, its ineffective extension across time and space was an insufficient trade-off dividend for its loss of reality. Phonetic writing, on the other hand, while an even poorer replicator of reality, provided a vastly superior extension across biological limits: indeed, Cyrus Gordon, Gerhard Herm, and similar students of the Ancient Middle East suggest that phonetic writing was invented by Phoenicians or even earlier sea-traders who found pictographs simply too cumbersome for their numerous commercial transactions. [6] Thus, an important secondary principle of anthropotropic evolution suggests itself here, to wit: that "Stage B" technologies evolve towards maximum extension across time and space in complete disregard of their ability to replicate reality; a poor extender across time and space such as hieroglyphics may thus be succeeded by an even worse replicator of reality such as phonetic writing, provided that the new technology improves upon the extensional capabilities of the old technology.

4. Hans Jensen, in *Sign, Symbol, and Script*, 3d ed., rev. and enl., trans. George Unwin (New York: Putnam's, 1969), p. 51, describes the evolution of writing to greater abstraction, i.e., greater divergence from reality, as follows: "The difference between the means of representation, 'abbreviated,' stylized,figurative, and what they were supposed to represent gradually became greater and greater, so great finally that the bond linking form and meaning could break completely, and instead the phonetic equivalent of the picture gained the upper hand: the picture was now no longer, or no longer solely the representation of an object or of the *meaning* of the word naming it, but at the same time, or even solely, a colorless and meaningless fixing of a certain *sound* or *sound-complex*. This transition is known as 'PHONETICIZATION.'" (Italics and capitalization in original.) Jensen, pp. 57-61, then describes several discernible stages in this evolution, including: "word picture-writing (ideograms)," in which objects are represented by stylized pictures of the objects (e.g., a stick figure for "man"), and abstract concepts are represented by pictures of

objects that are in some direct, visible way associated with the concept (e.g., a picture of a scepter meaning "to rule"); "phonetic word-script," in which concepts are represented by pictures of objects whose spoken words sound like the spoken word for the concept (e.g., a picture of a bee and a leaf meaning "belief"); "alphabetic script," in which objects and concepts are represented by combinations of individual units ("letters") which are not pictures of any objects (i.e., not the objects represented, nor objects associated with the concepts represented, nor objects whose spoken words sound like the spoken word for the concept represented), but are rather descriptions of individual sounds which combine to form the spoken word for the object or concept represented. As to appropriate terminology for the above stages, Jensen, p. 40, suggests that although the terms "hieroglyphics" and "pictographs" are often used interchangeably to describe all types of writing that entail pictures, the usage is a bit imprecise since some Egyptian hieroglyphs also entailed a phonetic process (e.g.,the "phonetic word-script"described above; indeed, Jensen, pp. 54-55, summarizes the suggestions of some scholars that hieroglyphics contained the rudiments of an alphabetic script); moreover, the existence of phonetic elements in hieroglyphics raises a comparable problem in the blanket equation of "phonetic" with "alphabetic" writing. Since, however, the present study is primarily concerned with the relationship of media to pre-technological reality, and since any use of pictures (even when combined with phonetics, as in some hieroglyphics) always bears a greater connection to reality than the "purely" phonetic (i.e., non-pictoral) alphabet, the present work will use the terms "hieroglyphics" and "pictographs" to refer to all types of picture writing, and the term "phonetic writing" to describe the completely non-pictoral phonetic writing of alphabets.

5. The slowness of picture writing would make it a poorer extender across space than time, where permanency is obviously more important than speed. This adds another dimension of support to Innis' contention that Egyptian and other early civilizations were timer-oriented, and, in general, suggests that time-oriented societies are maintained by any combination of three types of media: oral communication, writing upon nonportable materials, and writing in hieroglyphic or other pictoral

representations. See the end of Ch. 3 , including note 18, above.

6."The cumbersome scripts of Egypt, Mesopotamia, and China," Gordon writes in *Before Columbus* (New York: Crown, 1971), p. 94, "were all right for established, land-based cultures capable of maintaining an elaborate school system with many years of instruction before a scribe could qualify for his profession. Sea People needed some literate personnel for every operational unit — which happened to be the individual ship. Since vessels were by modern standards quite small, a system of writing that could be learned quickly was required. The alphabet — usually ranging between twenty and thirty letters — made sense, and it is no accident that a Sea People, the Phoenicians, employed the alphabet and transmitted it to Europe." Note that Gordon here credits the Phoenicians, "a Sea People," only with the employment and transmission, not the invention, of the alphabet. This reflects Gordon's view that the alphabet was in fact invented by a pre-Phoenician group (of whom the Phoenicians were derivative) whom Gordon calls "The Sea Peoples," and who, he suggests,also developed mathematics, astronomy, architecture, principles of economics, law, religion, techniques of agriculture, ceramics, and metallurgy — in short, much of what we recognize as the foundations of civilization — as they criss-crossed both sides of the Atlantic in their loose, maritime, empire (p. 105). The alphabet, however, was the lynchpin of their civilization, for Gordon believes that it served not only a phonetic but a mathematic and calendric function as well (this would be another example of the primitive emphasis of function rather than replication of reality in communication). Gordon credits much of this speculation to Hugh A. Moran and David H. Kelly, *The Alphabet and the Ancient Calendar Signs* (Palo Alto, Cal.: Daily, 1969); for more on the Phoenicians and their role in the development of the alphabet, see Gerhard Herm, The Phoenicians, trans. Caroline Hillier (New York: William Morrow, 1975), pp. 176-180.

The operation of this principle, as will be seen shortly below, was repeated at least once in the evolution of media, when print was in part succeeded by the telegraph, which provided a poorer replication of reality (dots and dashes

are abstractions of printed words, which are themselves abstractions of reality), but offered a greatly quickened extension across space (messages traverse the telegraph instantly, as opposed to the slower process of print). (It also follows from this principle that once "Stage C" technologies begin to appear — i.e., once technologies start to increasingly retrieve the pre-technological environment — the maximized extension of "Stage B" will continue. As will be seen more fully below, the evolution from radio to television, from black-and-white to color, etc., without any extensional loss supports this assumption.)

Regardless of their differences, however, hieroglyphics and phonetics share the overwhelming common identity of all written communication as examples par excellence of "Stage B" technology: both extend across time and space at the extreme sacrifice of the real world (hieroglyphics less so in both cases than phonetics).This "voice of an absent person," as Freud so aptly dubbed it (see beginning of Chapter 2 above), substitutes for the voice and the memory of it, and therein can project these across miles and years; and yet the voice of writing always arrives minus its original speaker, with the memory denatured of its original context. McLuhan and others, as noted above (Chapter 2), have explored many of the distortive consequences, of this sacrifice (especially in phonetic writing); and yet to the extent that so much of our civilization has been made possible by writing— helping to make "objective science" feasible, and enhancing the continuity of religion and art— the "sacrifice" certainly seems to have been eminently worthwhile.

PRINT: AT THE TURNING POINT

And what impact did the introduction of the printing press have upon this distortive though civilizing process? As has been seen earlier, McLuhan and others primarily view print as a magnifier of phonetic effects; simply put, print brought writing into a greater number of people's lives — an immensely greater number of people — thus promoting more separation of intellect from emotion, more abstractions of reality, more science, in sum, more of everything that writing engenders. And as Innis was the first to fully surmise (see Chapter 2 above), the magnified centrifugal force of print engendered commensurately magnified consequences, rending the Church in two and the feudal system completely apart, and stimulating such fragmentizing forces as privacy, individualism, and capitalism, as the fragmentation of reality through print found social expression. [7] Thus, print may be considered the very apex of "Stage B" technology— extending the spatial and temporal benefits of writing to an exponentially increased population, yet therein spreading the distortions of writing in equal proportion.

7. "Publication of the scriptures in the vernacular was followed by new interpretations and by the intensive controversies conducted in pamphlets and sheets which ended in the establishment of Protestantism. Biblical literalism became the mother of heresy and sects," Innis writes in *The Bias of Communication*, p. 54, adding that

"the printing press became 'a battering-ram to bring abbeys and castles crashing to the ground,'" p. 55. (The quote taken from G. M. Trevelyan's *English Social History*.)

But the very magnitude of the spread of privacy and separation through print was surely something of a two-edged sword: for when private circumstances are shared by large numbers of people, even not simultaneously, the privacy in some fundamental way must surely be diminished. In the pre-technological environment, communication is neither abstract nor particularly private; actual events are experienced by all within biological proximity, and cultural transmission, as McLuhan was early to recognize, consists of a common body of oral fable and saga shared by all who have ears. The introduction of writing, as has been seen above, shattered this unity first of all by facilitating and lengthening the separation of an event's description from its actual occurrence; but in addition to this primary separation, writing further shattered the unity of cultural transmission by making it in large part the exclusive province of the limited few who could read and gain access to scarce manuscripts. (Such a situation engendered the "castes of information" and "monopolies of knowledge," so termed and documented by Innis.)[8] The "abstraction" of writing can therefore be seen to consist of two abstracting effects: a primary, or "psychological" abstraction in which description is severed from reality, and a secondary or "social" abstraction in which a small literate elite is singled out as communicators. It can also be readily seen that print, as a magnifier of writing, only magnifies the first abstractional effect of writing — while in fact working to all but eliminate the second. Thus, while print, as detailed in the paragraph immediately above, no doubt made written abstractions of reality available to more people, this very process tended to undermine the monopolistically private concentration of information that hand-written communication made possible. The workings of this dual process are perhaps best epitomized in the very fissure of the Church, mentioned above and noted by Innis: in directly placing the abstractions of the Bible into more people's lives, the printing press severely undercut the role of the Church as the sole arbiter of the Bible's abstractions. Another consequence of the new sharing of abstractions, also noted by Innis, was the growth of national identities and states, stoked by the common self-perceptions that readers found in newly available literature and political tracts.[9]

8. See, for example, *The Bias of Communication*, p. 53, where Innis points out that "monopolies of knowledge controlled by monasteries were followed by monopolies of knowledge controlled by copyist guilds in the large cities."

9. Ibid., p. 55: "By the end of the sixteenth century the flexibility of the alphabet and printing had contributed to the growth of diverse vernacular literatures and had provided a basis for divisive nationalism in Europe," Innis writes.

Thus, while writing is both psychologically and socially fragmentizing, print seems to be psychologically fragmentizing but socially cohesive; and in helping to restore some of the social unity destroyed by writing, print obviously does more than merely magnify a "Stage B" process: indeed, its socially restorative powers suggest the very beginnings of a "Stage C" technology. Such a possibility was recognized by Innis, who pointed out that "mechanization facilitated an approximation of the printed word to the oral tradition."[10] An even more exciting discovery is that print's real-world retrieval was understood by its early practitioners, as indicated by 16th century artist Johannes Stradanus' caption to his engraving of a print shop's interior, which reads "Just as one voice can be heard by a multitude of ears, so single writings cover a thousand sheets."[11] If, as McLuhan suggests, "the phonetic alphabet translates man from the magical world of the ear to the neutral visual world" (see end of Chapter 2 above), then the promotion of the phonetic alphabet in print surely returns some of that lost acoustic magic — at least in so far as endowing the the abstract written word with an omnipresence akin to its oral counterpart.

10. *Empire and Communications*, p. 148.

11. Reprinted in Joseph Agassi, *The Continuing Revolution* (New York: McGraw-Hill, 1968), p. 26.

Moreover, Stradanus' engraving points to quite another, often unnoticed, way that print broke the monopoly of writing and helped retrieve the real world: the printing press, it is often forgotten, prints not only words but pictures, and pictures are usually less abstract or more literal approximations of the real world than words. Thus, in making engravings and woodcuttings such as those of Albrecht Dürer available to large numbers of people, the printing press not only retrieved a pre-technological, "ear-like" access to abundant information, but helped retrieve the real world in the information itself. It is perhaps no accident that the content and execution of these early engravings usually matched the pre-technological realism of their distribution; and art historian Erwin O. Christensen describes Dürer as "sober and factual," with a passion for detailing "every wrinkle, hair,and vein"[12] of his subjects. Indeed, in promoting both the creation and distribution of these precise depictions of reality, the printing press may be considered an important precursor of photography.

12.The History of Western Art (New York: Mentor, 1959), p. 241.

Thus, the printing press plays the role of a pivotal double agent in the evolution of media: greatly spreading the non-reality of writing, on the one hand, yet helping to retrieve the pre-technological situation of sharing fay the very process of spreading^ and retrieving both the situation and the content of pre-technology in its dissemination of pictures. Thus print, at the sheer physical summit of "Stage B" technology, contained in it the seeds of a new anthropotropic order.

TELEGRAPH AND PHOTOGRAPHY: THE TWO GREAT BRANCHES

The seeds remained seeds for nearly 400 years until the first quarter of the 19th century, when they suddenly found expression in the two radically new technologies of telegraphy and photography. The telegraph was of course the first technology to use electricity for instantaneous communication across huge distances — and although this process, as will be explained shortly below, was in many ways less anthropotropic than its predecessor the printing press, the first electronic medium set in motion a line of evolution that has resulted in transmission of ever more life-like messages. The photo-chemical process of photography, on the other hand, was recognizable in its inception as an unprecedentedly literal recording of the real world, by-passing the mental abstractions of earlier recording media; photography, too, begat a series of technologies that captured more and more of the real world, though these media would in many ways be ultimately co-opted by the progeny of the telegraph.

The very appearance of the Romulus and Remus of "Stage C" technology at almost the same moment in history suggests a striving for an anthropotropic balance of space and time extension, discussed above in connection with Innis (end of Chapter 3). Print itself had in many ways achieved such a balance during its long unchallenged reign: while the primary impact of the printing press was surely the distribution of more information to more places, i.e., extension across space, books were not an unendurable product, and thus their very abundance served to increase the amount of information preserved, i.e., extended across time. The effect of telegraphy, of course, was to wildly increase the extension of print across space — to the point where the telegraph may be said to have totally "vanquished" space, since distances of 10, 100, or 1,000 miles are telegraphically traversed with nearly equal instantaneous speed. But telegrams are neither designed for durability nor possess, like books, the capacity for it; and thus the neonate electronic medium stood poised to completely upset the time/space applecart in favor of space extension. Only the equally immense preservational ability of the photograph could have provided a sufficient counter-balance, which, coming at just the appropriate time, it did. Thus, first perfected in the 1820s, commercially developed in the 30s and 40s, becoming part of every-day existence by the 50s and 60s, telegraphy and photography are an excellent example of two media that, in conjunction, maintained the human balance of time and space communication while exponentially extending it. (The combined impact of telegraph and photograph obviously tipped the scales in a visual direction— an imbalance which was shortly to be rectified when the telegraph begat the "talking" telegraph or telephone, as will be described later in this chapter.)

TELEGRAPH: RETRIEVAL OF THE PRE-TECHNOLOGICAL PROCESS

Of the two technologies, the telegraph, as has been suggested above, was by far the less anthropotropic; indeed, the telegraph in many ways seems to be one of the "extreme" cases of "Stage B" technology in which, as in phonetics, an advance in extensional ability is paid for with a reduction in replication of reality. The extensional advantage of the telegraph is of course its instantaneous transmission across small and large distances; yet the content of these transmissions is a departure from pre-technological communication perhaps greater than that of any technology, either before or after the telegraph. If the written or printed word, as McLuhan puts it, is "a double level of abstraction" in that "the written word is an abstraction of the spoken word which, in turn, is an abstraction from the holistic experience,"[13] then the electrical messages conveyed by telegraph are surely a "triple" level of abstraction beyond holistic or pre-technological experience, since the short and long clicks of Morse code are themselves abstractions of the written or printed word. Thus the telegraph appears to bear much the same relationship to its predecessor print, as the phonetic alphabet did to its predecessor hieroglyphics, in that both telegraph and alphabet quickened the pace of communications, while leaving an even greater part of reality behind.

13. Marshall McLuhan and R. K. Logan, "Alphabet, Mother of Invention," *et cetera* 34 (December 1977); 377.

But if the telegraph and phonetic writing share the dual "Stage B" characteristics of greater extension yet worsened replication of reality, they differ sharply as to the relative impact of these two characteristics. In the case of phonetics, already alluded to above, it was the quantum detachment from the real-world sensory milieu that yielded the most far-reaching consequences: for while the speed and flexibility of abstract letters no doubt greatly facilitated the spread of writing, it was the abstractness itself that catapulted the visual from the other senses, as McLuhan has amply explained, and prepared the psychological climate for Euclidean geometry, objective science, rational/ linear thought, and other bedrocks of Western civilization. (Psychologist Julian Jaynes has recently offered support for this view by suggesting that phonetic writing was in effect partially responsible for the origin and perpetuation of "consciousness."[14]) In the case of the telegraph, however, its electric speed and flexibility proved of far greater consequence than its artificially coded messages, and soon dwarfed these messages into merely an interesting footnote to anthropotropic evolution. Indeed, the instantaneous interchange of the telegraph proved one of the great anthropotropic leaps in history— allowing for the retrieval of at least two previously elusive elements of the real world, as will be described below— and this in spite of the fact that the information so conveyed in this interchange was triply removed from the real world.

14. "The importance of writing in the breakdown of the bicameral voices is tremendously important. What had to be spoken is now silent and carved upon a stone to be taken in visually," Jaynes writes in *The Origin of Consciousness in the Breakdown of the Bicameral Mind* (Boston: Houghton Mifflin, 1976), p. 302. Jaynes' general thesis is that humans have recently evolved (beginning approximately 3000 years ago) from a state of dual mentalities, in which the left side of the brain literally heard "voices" emanating from the right side, to the integrated mental state that we call "consciousness," in which the inner voices have fallen silent. Although Jaynes often fails to distinguish between the parts played by pictographic and phonetic writing in the silencing of the voices, the decisive role of phonetic writing in this regard is obvious for at least three reasons: (1) while pictographs may have competed with inner voices and therein weakened their authority, phonetic writing in effect mounted a full-scale frontal assault on the voices by literally translating sounds into visual symbols; (2) the greater simplicity of phonetic writing made it in any case a more effective competitor of the voices; (3) the historical cases used by Jaynes as examples of the breakdown of the bicameral mind for the most part involve phonetic rather than pictographic writing (e.g., the example of Moses, from which the above quote is taken; the breakdown of the Greek bicameral mind as evidenced in the change of style from the *Iliad* to the *Odyssey*; etc.). For more on Jaynes' hypothesis, see James Morriss' "Comments and Reflections on a Theory of Linguistic Consciousness," *et cetera* 35 (September 1978), forthcoming; see also Ch. 6, n. 20 below.

In this sense, the telegraph can be seen to occupy much the same position in the evolution of media as the printing press, which, it will be recalled, promoted a detachment from reality by widely distributing abstracted (phonetic) information, yet retrieved the important real-world situation of sharing by its very distribution of information. By exponentially heightening this distribution factor, the telegraph further retrieved the real-world in two unprecedented, interrelated ways:

The instantaneous transmission of the telegraph is of course obvious as a great extension of information across space; perhaps not as obvious, however, is that this speed of transmission allows a condition of simultaneity across vast distances — recreating the simultaneous access to information that, as indicated in the previous discussions of McLuhan (end of Chapter 2) and print (earlier in the current Chapter), forms such an important part of the pre-technological communication environment. When information communicated via telegraph becomes almost instantaneously available in Baltimore, Washington, D.C., and New York City, and then, with the opening of transatlantic cables, in London and Paris, then the world surely becomes somewhat of a "global village," to use McLuhan's overused but still apt term, with New Yorkers and Londoners enjoying the same access to a single source of information as everyone within ear-shot of the town crier. That the abstract-bearing and in many ways "linear" telegraph was the first catalyst of this "global village" was recognized by McLuhan, who recently pointed out that the "electronic and simultaneous" situation "is not new, having begun with the telegraph years ago."[15]

15. "The Hemispheres and the Media," Centre for Culture and Technology, University of Toronto, Toronto, Canada, 12 February 1977, p. 14.

Generally overlooked by McLuhan and others, however, is a second, at least equally momentous anthropotropic consequence of the telegraph which, in making information simultaneously available across large distances, also allowed for a simultaneous exchange or interaction of information across space. In the pre-technological world, communication can either be "interactional" (two-way), as when two people converse and both send as well as receive information, or "observational" (one-way), as when a man gazes at the sunset and can only receive information without influencing its source. (Of course, it is a characteristic of the real world that many observational situations, such as eavesdropping on the conversation in the next room, can be transformed into interactive environments, in this example by merely entering the next room.) Until the telegraph, not a single technological medium permitted such an immediate interchange of information — in fact, primitive technologies to the contrary were able to extend across space and time only by imposing such delays as to render any interactive communication impossible, or limit it, in effect, to observation only. Even in the case of correspondence through letter-writing, where the intent is interactive, the effect is nonetheless one of strict sequential observation rather than interaction, with each party responding to the other in segregated absentia, or minus the common arena of communication that allows both parties to receive and send information simultaneously or in such rapid succession as to be in effect simultaneous. By allowing anyone hooked into the system to immediately interact with anyone else hooked into the system, the telegraph thus broke once and for all the observational yoke that had been the unavoidable burden of all previous technological extension across time and space.

With the pattern of real-world interaction thus re-established in technology through the structure of the telegraph, it was a comparatively simple matter to render the abstract content or messages of the telegraph in a more realistic form as well. The anomaly of Morse code was shortly removed with the invention of the telephone —literally a "talking telegraph,"[16] which, in utilizing the instantaneous, interactive structure of the telegraph to transmit human voices rather than dots and dashes, at once dropped the triple level of abstraction of the telegraph back to the single level of abstraction expressed in human speech. (And indeed, even this single level of abstraction would be somewhat removed by the most recent descendent of the telegraph — the video-phone, which interactively transmits real-world images as well as voices, see further discussion in Chapter 7 below.) With the "singing wire" thus transformed into the "talking wire," humans for the first time were able, at least so far as extension across space was concerned, to have some of their cake in communication and eat it too: people could now converse well beyond their biological or real-world limitations, without having to totally sacrifice all the useful and comfortable accoutrements of their real-world environments. That this retrieval of the real world was also interactive would prove of enormous consequence in the subsequent evolution of media — thus making the story of the telegraph, telephone, and video-phone one worthy of further consideration, both later in this chapter and in Chapter 7. For the present, however, it is enough to recognize that the telegraph, despite its "flaw" of abstraction, serves as the structural archetype for one of the most significant branches of anthropotropic evolution.

16. According to S. H. Hogarth, "Three Great Mistakes," *Blue Bell*, November 1926, this was the first name that Alexander Graham Bell used for his invention.

PHOTOGRAPHY: RETRIEVAL OF THE PRE-TECHNOLOGICAL CONTENT

The other archetype of anthropotropic media, the photograph, contained no such flaw, recapturing from its inception an essential content of the real world. Although the public reaction most usually associated with early photography is French artist Paul Delaroche's exclamation that "from today, painting is dead"[17] (failing to make McLuhan's observation that outmoded media rarely die, but spring instead, Phoenix-like, into new art forms), [18] *New York Herald* editor Horace Greeley was much closer to the anthropotropic mark with his observation that "in daguerreotypes we beat the world."[19] Unlike even the most "realistic" of paintings, which present images of the real world inevitably filtered through the more or less abstracting mentality of the painter, even the most carefully contrived photographs faithfully record an aspect of the world which is, to a greater or lesser degree, direct and undistorted. This unprecedented chasteness of the photographic image is recognized, as Stanley Milgram has recently pointed out, in the very language we use to describe photography. Milgram explains,

> The English language is blunt about the nature of photography. A photographer *takes* a picture. He does not *create* it. [Italics in original.][20]

Susan Sontag elaborates on the distinction in *On Photography*, explaining that

> [photographic] images are indeed able to usurp reality because first of all a photograph is not only an image (as a painting is an image), an interpretation of the real; it is also a trace, something directly stenciled off the real, like a footprint or a death mask. While a painting, even one that meets photographic standards of resemblance, never does more than state an interpretation, a photograph never does less than register an emanation (light waves reflected by objects) — a material vestige of its subject in a way that no painting can be.[21]

Sontag's analogy of the "death mask" is supported by the gruesome testimony of surviving early photographs — an inordinately large percentage of which were used to capture the images of the dead, especially children,[22] as if in a desperate attempt to hold on to some "material vestige" of a departed loved one.

17. This in reaction to L. J. M. Daguerre's showing of his "daguerreotypes" at a joint meeting of the French Academy of Sciences and Academy of Fine Arts on August

19, 1839. Daguerreotypes were the first commercially successful photographs, but unlike more modern photographs provided single, unreproduceable "positives" that developed directly on the original plate. See T. K. Derry and Trevor I. Williams, *A Short History of Technology* (New York: Oxford Univ. Press, 1961), pp. 653-654. Another distinguishing feature of daguerreotypes is that, unlike modern photographs, their surface was so bright as to serve as an actual mirror for any who beheld them — this making the first photographs literally as well as figuratively the "mirror of man." (For a brief description of the daguerreotype's "mirror-bright" picture, see Rita Reif, "Collectors Focus on Daguerreotypes," *New York Times*, 9 October 1977* sec. 2, p. 34.)

18. See Ch. 1, n. 27 above. Thus painting, unable to compete with photography as a mirror of the real world, survived by becoming more abstract and impressionistic — a development which might be viewed as an enhancement, rather than a destruction, of painting's art.

19. Cited by Beaumont Newhall, *The Daguerreotype in America* (n.p.: Duell, Sloan & Pearce, 1961), p. 11.

20. "The Image-Freezing Machine," *Psychology Today* 10 (January 1977): 52.

21. (New York: Farrar, Straus, and Giroux, 1977), p. 154. It must be noted that Sontag in no way regards the photographic approach to reality as a benefit to humanity — suggesting instead, on p. 24, that we have become a nation of "image-junkies," who prefer the easy, but ultimately unreal substitute of photographic reality to the real thing. But see "Appropriate Monsters" in Ch. 9 below for a rejoinder to this viewpoint.

22. "Reif, "Collectors Focus on Daguerreotypes."

That a machine could be pressed into such humanistic service is the crux of the anthropotropic emergence. With the camera lens enjoying the same unmediated relationship with the real world as the human eye, photography became the first equivalent of human memory — rather than a codified record or abstraction of it — and could thus serve in memory's stead as a functioning organ of human life. Improving upon both the specificity of writing and the graphic expanse of painting in one fell swoop, photography in many ways functioned as the ideal of memory, being the first medium in history — technological or human

— to extend the real world across time without totally transfiguring it. (Of course, abstract writing still offered the most undistorted recording of abstract ideas, see Chapter 6 below.)

Thus, the photograph heralds the advent of a new type of technology which, when set in motion by humans, paradoxically achieves a human effect in some ways greater than that achieved by humans acting alone. The full dimensions and significance of this palpable turn of events — as well as its fulfillment of fundamental human desires — were perhaps best understood and summarized by André Bazin:

> For the first time, between the originating object and its reproduction there intervenes only the instrumentality of a nonliving agent. For the first time an image of the world is formed automatically, without the creative intervention of man.
>
> The personality of the photographer enters the proceedings only in his selection of the object to be photographed and by way of the purpose he has in mind. Although the final result may reflect something of his personality, this does not play the same role as is played by that of the painter. All the arts are based on the presence of man, only photography derives an advantage from his absence....
>
> . . . Only a photographic lens can give us the kind of image of the object that is capable of satisfying the deep need man has to substitute for it something more than a mere approximation, a kind of decal or transfer. The photographic image is the object itself, the

object freed from the conditions of time and space that govern it. No matter how fuzzy, distorted, or discolored, no matter how lacking in documentary value the image may be, it shares, by virtue of the very process of its becoming, the being of the model of which it is the reproduction; it *is* the model. [Italics in original.]

Hence the charm of family albums. Those grey or sepia shadows, phantomlike and almost undecipherable, are no longer traditional family portraits but rather the disturbing presence of lives halted at a set moment in their duration, freed from their destiny; not, however, by the prestige of art but by the power of an impassive mechanical process; for photography does not create eternity, as art does, it embalms time, rescuing it simply from its proper corruption. [23]

Of course, there were many aspects of the real world that early photography failed to record. One limitation of the daguerreotypic process that served as the first commercial photography was that daguerreotypes were single, unreproduceable prints, thus undermining the new tribal cohesiveness promoted first by the printing press and then the telegraph. This shortcoming was soon remedied by Britisher Fox Talbot,- who produced the first photographic "negative," capable, in theory at least, of producing copies m mass. [24]

23. *What is Cinema?*, pp. 13-14.

24. According to Derry and Williams, pp. 655-656, Fox Talbot's technique could print hundreds of copies within a week; this was improved upon in 1851 by the Frenchman Blanquart-Everard, who introduced a process

that could make hundreds of photographic copies in a single day.

MOTION PHOTOGRAPHY: BEYOND THE LANDSCAPES OF THEATER

A more profound (and more difficult to rectify) limitation of early photography was its inability to capture motion. Attempts to technologically replicate the motion of the real world were initially conducted independently of early photography, and were evidenced in the various "wheel of life" devices of the 1820s and after, which, utilizing the persistence-of-vision phenomenon noticed since Ptolemy, simulated real-world motion by presenting a series of sketches in rapid succession.[25] Not until the wheel of life's union with the actuality of the photographic image, however, would the device fulfill the promise of its name, and convey the motion of the living — rather than the depicted — world. The more or less common development of such a motion photography process by Edison in America, Freize-Greene in England, and various Frenchmen towards the end of the 19th century suggests the seeming inevitability of anthropotropic progress once the necessary ingredients are available. (Indeed, this pattern of parallel and often nearly simultaneous invention characterized most of the communication developments of the late 19th and early 20th centuries.) [26]

25. See Mast, pp. 18-22, for a brief discussion of the "Thaumatrope," "Phenakistiscope," "Stroboscope," and other devices of this sort.

26. McLuhan, for example, points out in *Understanding Media*, p. 237, that "the American Patent Office received Elisha Gray's design for the telephone on the same day as Bell's, but an hour or two later. . . . Bell got the fame, and his rivals became footnotes." Other examples of such independent and nearly simultaneous invention are described by Derry and Williams: telegraphy (Schilling and Morse), pp. 624-626; photography (Niepce/Daguerre and Fox Talbot), pp. 652-653; and, much earlier, the printing press (Gutenberg, Fust, and Coster), p. 238. Of course, as the communication process itself improved through these inventions, instances of genuinely independent discovery might be expected to become rarer. Nevertheless, Erik Barnouw has written of David Sarnoff's attempt to "coordinate all moves" in the development of television: "the veteran patent fighter was faced once again with patent problems — not from rival corporations, but from inventors in the mold of the boy Marconi inventor of the radio — inventors who insisted on inventing on their own." *The Tube of Plenty* (New York: Oxford Univ. Press, 1975) , p. 77.

As a successor to still photography, motion photography exemplifies the workings of "Stage C" or anthropotropic evolution: maintaining all the extensional abilities of still photography, motion photography retrieves all that still images do, in addition to retrieving something — the motion of the real world — that still photographs fail to retrieve. (Bazin defines still photography as the preservation of objects, and motion photography as the preservation of objects in their "duration" or "change.")[27] In other words, unlike the taking-from-Peter-to-pay-Paul syndrome that characterizes much of the evolution of primitive media, motion photography provides an uncompromised "improvement" over still photography.

27. *What is Cinema?*, pp. 14-15.

Yet in improving upon still photography to the point of substituting, not only for the memory of images, scenes, and isolated incidents, but for the mental ability to recall or construct entire stories of connected events, motion photography becomes a successor, not only of still photography, but of the traditional narrative media of theater and the novel as well. Indeed, performing in the shadow of the theater until the First World War, and easily overshadowing theater afterwards, motion photography seems in many ways more a successor of theater than of still photography (indeed still photography, for good anthropotropic reasons, has survived the advent of motion photography a lot better than theater has, see discussion in Chapter 6 below). It thus becomes necessary to ask of motion photography: in what way does it improve upon the narrative reality of theater? Or, put less charitably to the present thesis, doesn't film in fact move a step away from the reality of live theater? (Of the relationship between film and the novel, there can be little argument: film improves upon the novel's ability to transmit pre-technological concrete environments in the same way that a still photograph improves upon the word.)

To the extent that a film is only a motion picture of a live theatrical presentation — and most early films were — the answer to the above question would per force be yes. Certainly a film of a prop that stands for a mountain, of an immense battle scene that stretches 10 feet across a stage, of actors and actresses dressed in exaggerated costumes and make-up so as to be seen in the back rows — certainly, a film of these theatrical staples would be further removed from the real world than the actual theatrical presentation itself.

But film has long since left the proscenium arch for the

wide-open — or at times more intimate — spaces of the real world, where protagonists climb real mountains, where rivers flow with real water rather than the water-colors of backdrops, and where people's faces express the subtle nuances of life rather than the caricatures of the stage. In effect, the environment of film-making allows for a highly faithful recording of living structures and relationships, while the environment of theater, though offering life itself rather than a recording of it (in the case at least of actors and actresses), often forces this life to assume highly unnatural and distortive postures. The celluloid images of life are thus in many ways more life-like than the conceits and contrivances of living theater.

The discussion at the close of the last chapter about the multiple viewing perspectives of film may also have some relevance to a comparison of film and theater. While the theater audience is limited to the single perspective of whatever is facing them upon the stage, the film audience enjoys the multiple perspectives of numerous camera angles, shifting (albeit at the will of the film editor) to the sides and behind, above and below, to close as well as distant relationships to the action of the narrative. Since perception in the real world is obviously a continuing mixture of such shifting perspectives and distances, the repertoire of film shots that comprise any movie must have greater consonance to the rhythms of life than the frozen point of view of theater.

Moreover, film improves upon the "reality" of theater in yet another dimension. As a medium for propagating the sharing of information across large geographical distances, theater is only slightly better than the hand-written manuscripts that pre-dated print. However, motion photography — which from its inception utilized a process of negatives capable of producing numerous copies —

continued.in the tradition of the printing press, which, while no match for the electronic simultaneity of the telegraph and its offspring, nonetheless made its information commonly available to ever-larger communities of discourse. And in the case of film, this pre-technological reality of shared information was enhanced by the realistic form of the information itself — which of course was composed of literal photographs rather than the abstract letters of print.

Thus, with the possible exception of theater's use of live performers — mitigated, as has been seen above, by the often distortive attitudes that these performers are forced to assume — film both, socially and perceptually seems to replicate more of the real world than theater. Motion photography thus seems a sizeable anthropotropic advancement, not only over its still photographic derivation, but upon its theatrical competition.[28]

28. Monaco, pp. 34-37, points out that theater, as might be expected, survived the challenge of film by emphasizing what in effect is theater's only anthropotropic advantage over film: the ability of theatrical performers to interact with audiences. Thus, Monaco cites Bertolt Brecht and Antonin Artaud, "two very different theorists of theater" who nonetheless "developed concepts of theater that depended upon the continuing interaction between audience and cast." Advertisements aired on N.Y.C. television in the summer of 1978 for the Broadway theatrical production of *Grease*, designed to respond to the competition of the film production of *Grease* released at that time, emphasize theater's interactive advantage over film by advising viewers to "See *Grease* the way it should be — live — where you can reach out and touch it." And an example of theater exercising this anthropotropic superiority over film in a slightly different dimension comes from the 1977/1978 Broadway production of *Dracula*, in which a musty powder is blown into the audience so as to simulate the atmosphere of a crypt. (It is interesting to note here that whereas painting withstood the advent of photography by emphasizing the abstract, theater met the challenge of film by

emphasizing the real. This no doubt reflects the technological capacity of film for both greater reality, in most cases, and greater abstraction than theater — with live interaction between performers and audience being just about the only advantage, real or abstract, that theater has over film.)

Of course, one area in which theater was initially a good deal closer to real life than motion photography was theater's presentation of sounds synchronized with sight. This limitation of "silent" film — which was not really silent but speechless, i.e., early films lacked the precise sight-and-sound synchronization needed to present speakers, but were often accompanied by background music — would soon be rectified by a technology that could coordinate the recording and playback of sounds with images, thus making speaking faces feasible. Ironically, the seeds of this new synchronized sound technology lay in the phonograph, perfected almost 15 years before the first motion photography, and, in the case of Thomas Edison, the very spur for the invention of motion photography itself! [29]

29. Josephson, p. 385; Mast, pp. 25-26. Monaco, p. 56, finds an even greater irony in the fact that Edison's phonograph was mechanical rather than electric, and wonders "whether there would have been any period of silent cinema at all had Edison not invented a mechanical phonograph: in that case it's quite possible that Edison (or another inventor) would have turned to Bell's telephone as a model for the phonograph and the electrical system for recording sound would have developed much earlier, more than likely in time to be of service to the first cinematographers." Monaco, however, overlooks the point — to be discussed more fully in the text immediately below — that film itself is a mechanical rather than electric process, and that the first sound synchronized movies recorded sound on film rather than electronic tapes. See note 33 below in this chapter.

PHONOGRAPH: "PHOTOGRAPHING" THE SOUNDS OF LIFE

The phonograph was actually not the first technology to capture the reality of the human voice — Bell's telephone in 1876 predated Edison's phonograph by a year — but as a mechanical rather than electronic device that sought primarily to extend a part of reality through time, the phonograph is very much a part of the photographic branch of anthropotropic evolution. Like still and motion photography — and unlike the telegraph — the phonograph from its very beginning retrieved a literal piece of the real environment, and sought to preserve it across time (the thrust of early electronic technologies like the telegraph and telephone was of course extension across space). Introduced, moreover, in an age in which advertising had already become a main mode of entry for new technologies into society, the phonograph was the subject of numerous advertisements which were abundantly aware of its anthropotropic attributes — and used these as the phonograph's prime selling point. Thus, an 1899 ad for Columbia's "Gramaphone Grand" boasts, perhaps a bit prematurely, of "reproductions so perfect that the ear cannot distinguish them from the or i g i n a l."[30] By 1917, both the rhetoric and the technology of the phonograph had been refined so that Columbia could write the following ad for its "Grafonola," worth repeating here in its un-edited entirety:

> The record played on the Columbia Grafonola is more than a record — it is *reality*. Through the marvelous Columbia reproducer, every individual musical pulsation — every modulation of every note comes back with volume and *warmth* the same as the very original itself.
>
> Its reproduction is as true as a mirror to every beauty of musical art — a triumph of perfected scientific precision. The splendid *resonance* so essential to reproducing orchestral music; the *delicacy* needed to carry the notes of whispering woodwinds and murmuring strings; the ability to convey the *living warmth* that gives great voices their personality — these make up the miraculous perfection of the Columbia reproducer and Columbia TONE.
>
> Clear, natural, brilliant, true — these words are hardly enough to describe it. Only *one* word can truly tell all that 'Columbia tone' implies — and that single word is: LIFE! [Italics and capitalization as in original.] [31]

Emphasis on the phonograph's ability to retrieve the voices and sounds of the real world continues in advertising to the present day, in which Pioneer Systems, to cite but one example, uses the slogan "we bring it back alive" to describe the performance of its musical reproductive systems. [32] Of course, in the case of Pioneer, its claims carry the authority of advances in technology which allow for the reproduction of sound in vast ranges of high and low tonal nuances, that approach the listener from a 360 degree total surround — thus offering a much closer approximation of the sound environment of the real world, in which sounds can and do approach the listener from all angles in the 360-degree horizon, than could possibly have been provided by Columbia's "Grafonola" or any of the phonographs which until the 1960s presented sound from a monaural or single perspective source (see Chapter 6 below for a discussion of monaural vs. multi-dimensional sound).

30. "The Gramaphone Grand," advertisement for Columbia Phonograph Co., New York, 1899.

31. "Columbia Grafonola," advertisement appearing in *The Delineator*, October 1917, p. 41.

32. Advertisement broadcast on WNBC-TV, 9 October 1977, 12:14 AM; another Pioneer ad broadcast on WNBC-TV at 12:35 AM on April 23, 1977 shows a picture of a caveman with the narration, "If man had a Pioneer cassette, he could take along the quality of the live performance, without the live performance." Meanwhile Scotch, a competitor, broadcast a series of ads in 1977-78 with the tag line: "With Scotch Brand Recording Tape, the truth comes out"; and Memorex Tape, in a series of ads broadcast in the same time period, defies audience and music stars to determine of music being played: "Is it live, or is it Memorex?" While the preponderance of these ads of course in no way "proves" that the claims made about retrieving reality are justified, the ads do demonstrate something of almost equal importance to the present anthropotropic theory: namely, that advertisers believe that retrieval of reality is what their customers most want.

But of course such advances only demonstrate the continuing evolution of media towards increasing replication of real-life communication environments. The lesson of Columbia's hyperbole in 1917 is that the phonograph was in its intention and growing performance as remarkably a replicative device as the photograph, and thus may be considered as great an anthropotropic leap in the world of sound as the photograph was in the world of

vision.

"TALKIES": REUNION OF SIGHT AND SOUND

As long, however, as photograph and phonograph continued to evolve separately, neither could fully retrieve a real world in which images and sounds are so integrally intermingled. The advent of "talkies," therefore, which in effect co-joined the phonograph with motion photography, was one of the most profound anthropotropic developments of the 20th century. That such a development was all but inevitable is borne out both by the frequent attempts, such as Thomas Edison's, to perfect a talking picture process long before its eventual introduction in the 1920s (indeed, as has been mentioned above, Edison is said to have invented his motion photography for the explicit purpose of providing a visual accompaniment to his phonograph; see also Bazin's observations on this subject, above), and by the swift and total demise of the "silent" movie once the "talkie" was introduced. [33] Both events — before and after the talkie, as it were — demonstrate that the desire to capture the coordinated sights and sounds of the real world was strong in both inventor and consumer. And this in turn reaffirms an observation first made in the above discussion of Freud (Ch. 2 above), namely that anthropotropic evolution is a nonarbitrary matter that proceeds in accordance with a persistent human desire to recreate in media as much of the real-world communication environment as possible.

33. "By 1929 the silent movie was dead in America," writes Mast, p. 230. See also Mast, pp. 225-229, for brief discussions of the "Vitaphone" (first American sound film process, which synchronized a recorded disc with a film projector), "Tri-Ergon Process" (earlier German technique that recorded sound directly on film), and the numerous attempts at synchronized sound movies that preceded these first two commercially successful processes.

Thus, in a century of evolution, the photographic process had advanced to retrieve both the motion and synchronized sounds of the real world (each achievement taking roughly 50 years to perfect). The retrieval of the next real-world element, color, would be less momentous but now more rapidly attained.

COLOR: LESS THAN LIFE, MORE THAN LIFE

The capturing of color, like synchronized sound and motion before it, had long been a goal if not an accomplishment of photography. George Melies and other filmmakers, as Bazin has already observed (see end of Chapter 4 above), painted their black-and-white films as early as the turn of the century to restore the missing dimension of color. In still photography, the practice of adding rose-coloring to the cheeks of subjects goes back at least as far as the daguerreotypes of the 1850s;[34] and in the first four decades of the 20th century, photographers such as Wallace Nutting carefully painted their snapshots of woodlands and colonial houses and sold them to a widespread middle class clientele.[35] The introduction of color as an integral part of the photographic process for both still and motion pictures in the late 1930s was thus a long anticipated event which removed the dimension of color from the abstract subjectivity of the painter, and placed it in the more direct, immediate sphere of the photographer. In effect, this transformation from painted color photography to photographed color photography may be considered a minor repetition of the seminal transformation of painting to photography itself.

34. Newhall, "A Quest for Color," pp. 96-110.

35. Marilyn Dipboye, "Wallace Nutting, Photographer," *American Collector*, November 1977, pp. 27, 30.

Much has been written and said about the "artificiality" of technicolor and other technological colorings,[36] and it is certainly apparent that the deco and dayglo colors of technology often attempt to surpass rather than replicate the colors of the real world. To these "criticisms" of technological color, two rejoinders can be made: (1) That the blue pigments, for example, that painters use to depict the blue sky seem no closer to the actual blueness of the sky than the colorings of photographic images. Indeed, a painter might very well prize an "unnatural" coloring as one way of transcending, rather than mimicking, reality — a technique which a photographer might also make use of in the pursuit of art rather than practical communication (such as a snapshot of a loved one). Thus, photographic color, on the one hand, seems no less realistic than painted color; while the popularity of unnatural technological colorings, on the other hand, might be explained as a deliberate effort to transcend reality or be artistic. (See discussion of art in Chapter 1, n. 30, and Chapter 4, n. 49 above.) (2) But what of the technologies which, unlike the 1960s psychedelic attempt to obviously surpass reality, are intended to capture the color of the real world as accurately as possible, yet, nonetheless, fall short? Such shortcomings can be accounted for in anthropotropic theory by the reminder that they are not by design — that is, failure to fully replicate the colors of the real world is an unintended consequence of technologies still too primitive to do the job. That there is a deep human desire for full color retrieval is demonstrated by the continuing ad campaigns for television sets that claim more "true-to-life" color in much the same terms as the early phonograph ads boasted of "living" sound.[37] (See below in this chapter for a discussion of other anthropotropic advances in television.)

36. Indeed, the term "technicolor" has been often used as a very metaphor for artificiality, as in Michael Franks' "Night Moves" song appearing on *The Art of Tea* album released by Warner Brothers Records in 1976, in which the singer describes love as "just another technicolor romance on the screen." But listen also to Paul Simon's "Kodachrome" song appearing on the *There Goes Rhymin' Simon* album released by Columbia Records in 1973, in which Simon uses color photography as a metaphor for imagination (or enhancement of reality through memory), and contrasts this favorably with the "black and white" of literal reality: "Kodachrome/ They give us those nice bright colors/ They give us the greens of summers/ Make you think all the world's a sunny day/ I got a Nikon camera/ I love to take photographs/ So mama don't take my Kodachrome away/ If you took all the girls I knew when I was single/ And brought them all together for one night/ I know they'd never match my sweet imagination/ And everything looks worse in black and white."

37. RCA, for example, fielded a series of ads in 1977-78 for its new "Colortrak" system, in which actresses describe subtle color nuances in their eyes, hair, and complexions, so as to impress the viewers with the accuracy of the color transmission. Similarly, Panasonic promoted its "Quintrix II" color tube in 1977-78 with huge billboards in the New York City area that depict footballs and other items jumping out of television screens into the living rooms of astonished viewers, with the caption: "Quintrix II from Panasonic. So lifelike you'll feel you're part of the picture."

ELECTRONIC CO-OPTION OF THE PHOTOGRAPHIC BRANCH: REUNION OF PROCESS AND CONTENT

The shift of the color debate from photography to television symbolizes a major shift in the course of anthropotropic evolution: for with the achievement of color replication, the mechanical, photo-chemical process of photography made its last successful contribution to the development of ever more life-like media. The next step in the retrieval of the real world would be the development of an image and sound process that was immediately accessible — as are the images and sounds of the real world, which don't have to wait hours for photochemical development and days for mechanical distribution. Television provided such an immediate audio-visual distribution; and its benefits were recognized as early as 1935 by Rudolf Arnheim, who wrote of the still infantile medium:

We see the citizens of a neighboring town assembled in the market square, the Prime Minister of a foreign country making a speech, two boxers fighting for the world championship in an arena across the ocean, the British dance bands performing, an Italian coloratura singer, a German professor, the smoldering remains of a wrecked railway train, the masked street crowds at the carnival, the snow-capped mountains of the Alps as they appear through clouds from an airplane, tropical fish through the windows of a submarine, the machines of a car factory, an explorer's ship battling the polar ice. We see the sun shining on Mount Vesuvius and, a second later, the neon lights that illuminate Broadway at the same time. . . . The wide world itself enters our room. Television . . . gives us a feeling for the multiplicity of what happens simultaneously in different places. For the first time in the history of man's striving for understanding, simultaneity can be experienced as such, not merely translated into a succession of time. Our slow bodies and nearsighted eyes no longer hamper us.[38]

38. "A Forecast of Television," originally published in February, 1935, and reprinted in Film As Art , p. 194. It is interesting to note that, much like his contemporary Mumford, Arnheim had grave misgivings about the evolution of media even as he glimpsed its outcome. Thus he warns of television, "although the new victory over time and space represents an impressive enrichment of the perceptual world, it also favors the cult of sensory stimulation . . . as we render man's image of his world immensely more complete and accurate than it was in the past, we also restrict the realm of the spoken and the written word and thereby the realm of thinking. The more perfect our means of direct experience, the more easily we are caught by the dangerous illusion that perceiving is tantamount to knowing and understanding." For rejoinders on the incompatibility of television and book knowledge, see "Television in the Vanguard" later in this chapter, and the concluding discussion in "The Persistence of Abstraction" in Chapter 6 below. See also "The Dream and the Act" in Chapter 9 below for a general consideration of the dangers of "direct experience" and how they might be dealt with.

But electronic television owed its speed and simultaneity, and lineage, to the telegraph/telephone line of media evolution, and not to the photograph. In a sense, then, the evolution from motion photography to television represents the triumph of the electronic over the photographic branch of anthropotropic media — a triumph which saw the photographic accomplishments of color, sound, motion, and image replication eventually subsumed into the content of a fundamentally telegraphic, or instantaneous, distribution system. (The triumph may be not all that complete, however, in that holography, an essentially photographic process, has recently replicated the previously unattainable third dimension in both still and moving images. See the end of the present chapter and "Future Reunions" — Chapter 7 — for further discussion of this as yet uncommercialized anthropotropic development.)

TELEPHONE AND RADIO: INTERACTION VS. ACCESS

Television, then, may be considered a direct descendent of the telegraph — but of a telegraph which had itself undergone considerable evolution. The invention of the telephone, it will be recalled, was significant in that it brought the content of the telegraph in line with its anthropotropic distribution system — that is, it made human voices rather than long and short clicks the content of an instantaneous delivery system that resembled the immediate access environment of the pre-technological world. But there was a serious flaw, from the anthropotropic perspective, in even the delivery system of the telegraph/telephone: for while the system was interactive and immediate, it was accessible only to those who had access to a fixed, stationary wire. But in the real world, information is immediately accessible to individuals who are mobile and not tied down to a fixed system. And in practical terms, wired transmissions meant that the information environment of the telephone could not be shared by people aboard sea, and later, aircraft.

The limitation was. rectified at the turn of the century with the invention of the radio, or "wireless." The distribution of information through electro-magnetic waves that traversed the air both freed communication from the restrictions of fixed systems, and came closer to approximating the pre-technological environment in which information is also distributed through the air. The development of portable transistor technologies later in the 20th century gave the radio listener further mobility and freedom from the unnaturally fixed systems of earlier technologies.

Radio was a genuine advance, moreover, in that it humanized the distribution system of the telephone, without sacrificing its human content or voice. But the elimination of the telephone's wires in radio was attained only with the sacrifice of another important anthropotropic feature of wired communication — namely, its capacity for interactive or two-way conversations. For radio to have retained the interactive capacity of telephone, radio transmitters would have had to have been installed along with every radio receiver — a situation which was economically unfeasible for all but a few "ham" operators, and, in the case of portable radio . receivers, all but technologically unfeasible for a very long time. The recent. "CB" or interactive Citizen's Band radio fad, indeed, is but the latest attempt to devise a portable as well as interactive communication system that has largely eluded economically feasible technology. Thus, on the anthropotropic score card, radio is not as clear an advance as is, for example, motion over still photography — in that radio improved upon the access of telephone, while losing its real world interactive ability. One consequence of this incomplete progress was to give fixed wire systems a monopoly in interactive communication (an advantage which, as will be discussed in Chapters 6 and 7 below, may account for the survival of fixed communication systems in a fluid world).

Thus, radio not only became the first electronic technology to instantaneously extend information to mobile receivers, but the first electronic technology to adopt an observational or one-way distribution mode. And since the entire photographic branch of media was obviously observational rather than interactional, radio was obviously an ideal distribution system for the content of motion photography.

Thus was television born: a radio delivery of moving pictures and sounds. The advance of television over radio was of course the addition of moving images to sounds; the advance of television over motion photography was the substitution of electronic for mechanical speed, permitting an instantaneous and simultaneous access to pictures and sounds more in accord with the immediate environment of the pre-technological world.

TELEVISION IN THE VANGUARD: LIFE-SIZE SCREENS, HOME VIDEO RECORDING, HOLOGRAPHY, FIBER OPTICS

As a replication of the pre-technological observational mode, television thus represents the current high water mark of anthropotropic evolution or retrieval of the real world, combining the proximity previously attained by photography to the content of the real world, with an electronic simulation of the process of real world communication. Moreover, television has itself been subject to continuing anthropotropic development. Its change from black-and-white to color images in the early 1960s (thus giving it almost complete content equivalence to motion photography) was but the most obvious of these adjustments; and more recent and still developing improvements in size, permanence, and access areas have helped heighten the precision, if not the scope, of television's retrieval of the real world.

Image size is a difficult standard upon which to judge technology against the real world, for in the real world we often correctly perceive an image of a man that takes up only a fraction of an inch to "really" be a man of "normal" size standing far off in the distance. [39] It is these size and distance constancies that allow us to view a person in a three inch photograph or on a twenty-inch TV screen, and compute the person as being a normal, 5'9" height. But yet there has always been something of an absolute, too, in the relationship of media size to reality, which explains why panoramic film has been a more satisfying medium for landscapes and long-shots, and small-screen television a more appropriate presenter of "talking heads" and close-ups. (The unsuitability of the galactic movie epic *Star Wars* to the small television screen was the source of a joke on a recent *Saturday Night Live* TV program;[40] the term "small screen," incidentally, is listed in the *New World Dictionary* as colloquial for television.)[41] In view of the above, the development of a system that both reduced the film image and increased the TV picture to life-size — i.e., to human proportions, or approximately five to six feet tall — would have to be regarded as an anthropotropic refinement that presented a more consistent and less symbolic replication of the real world, at least in so far as size is concerned.

39. Psychologists refer to this as size and distance "constancy," or the ability to perceive the sizes and positions of objects as we "know" them to be (e.g., the "normal" height of a man), regardless of the perspective, angle, or distance from which they are viewed. As Hilgard, pp. 191-192, has pointed out, however, we also have a contrary tendency to judge objects on their external perspective, such as seeing an object in the very far off distance as smaller than it actually is. Hilgard suggests, then, that size perception is often a "compromise" between our internal mental expectations (size constancy) and the physical perspective of the object (the varying size of the object's image on our retina).

40. The segment of this program, broadcast on WNBC-TV at 12:04 AM on April 2, 1978, features "star gazer" Bill Murray, who is making predictions as to what movies will win the up-coming Academy Awards. Murray informs us that he is not really in a position to say much about most of the movies nominated because he didn't see them, and as for *Star Wars*, he can't really make a prediction about that one either, because he saw it "on a television screen in a hotel up in Canada."

41. Second Collegiate edition, 1974, p. 1344.

The recent "Advent Video Beam" and other "wall size" television screens perform just such a service. Once again, the advertiser's copy tells a remarkably anthropotropic story — this time, about Advent's "Video Beam Television":

> To begin with, it's ten times the size of ordinary TV. But people who watch VideoBeam television don't talk about how big the picture is. They talk about how it gives the viewer the feeling of being there. The picture has a way of wrapping around and involving the viewer that is totally unlike the ordinary TV experience. It has a quality of total realism that is almost three dimensional. Things seem to jump right out of the screen. Actions and emotions have such immediacy they are literally felt by the viewer. There is a tactile quality. In other words, things are brought to life[42].

A large element of truth can once again be recognized in the advertiser's hyperbole: namely, that there is something in a six foot image that evokes increased consonance to the experiences of the real world.

42. Advertisement appearing in *The New York Times Magazine*, 4 December 1977, p. 155.

The question of television's permanency — or its ability to

extend across time— has been the source of much serious criticism of the medium. A balance between space and time extension in technology — first advocated by Innis (see end of Chapter 3 above), and suggested by the present theory as an inevitable reflection of the balance of space and time in human perception — was, it will be recalled, attained to a certain degree by print, and continued with the co-development of space-oriented telegraphy and time-oriented photography. The balancing act continued, as has also been already noted (at end of Chapter 3 above), with the invention of the time oriented phonograph only a year after the space-oriented telephone (indeed, the obtaining of a permanent record of telephone conversations is the main reason that Edison has given for his invention of the phonograph).[43] And similarly, the spatial dominance of radio in the 1930s was successfully countered by the temporally extensive motion picture industry, then at the height of its popularity and power. But the equilibrium was once again severely threatened by the advent of television — which eclipsed the motion picture as the predominant audio-visual medium by presenting images which were accessible to everyone simultaneously, but only for the explicit instant in which they were presented. At least the mechanical motion picture house allowed the viewer to sit through the movie a second time, or see it again later in the week.

43. Josephson, p. 161, reports that "Edison's friend Johnson has indicated that the inventor at the time was working on a commercial project: to record and reproduce sound coming over Bell's telephone. As he [Edison] described it to Johnson, 'it would be a telephone repeater — it would transmit, repeat, be of great practical value ....'" (Italics in original.) In other words, just as Bell was apparently aware that his telephone or "talking telegraph" provided an "ear-balance" to the telegraph (see n.16 in this chapter), so Edison was intentional in his phonograph's ability to time-balance the telephone.

It is the spectre of this extreme spatial bias of early television that has led critics such as Lewis Mumford to warn against "an electronically induced mass psychosis . . . the 'burning' of the books . . . nothing less than the erasure of man's diffused, multi-brained, collective memory. . . . . To be aware only of immediate stimuli and immediate sensations," Mumford adds, "is a medical indication of brain injury."[44] Mumford would thus presumably be relieved to learn that the once fleeting images of television are now retrievable by the general public through new home video-taping systems such as Sony's "Betamax" and RCA's "Selectavision"[45] (as indeed, radio broadcasts have been since the commercial introduction of audio-tape in the late 1950s, and the advent of purchasable phonograph records as the main content of radio broadcasts even before then). With the public now increasingly able to replay any television broadcast at will, the instantaneous electronic medium has become time as well as space binding — TV no longer "burns" books; it has, at least in terms of time extension, become books. Moreover, unlike earlier media environments in which time and space balance was achieved through a precarious cooperation of disparate, individual media (e.g., telegraph and photograph, telephone and phonograph, etc.), the time/space balance of television is all part of one media system — and thus in much closer consonance to the living balance of space and time perception in the one human system.

44. *The Pentagon of Power*, pp. 294, 298.

45. See David Lachenbruch, "The New Boom in Video-cassette Recorders," *House and Garden*, November 1977, pp. 64-65, for a brief description of current home video recording systems.

In the most recent — and as yet largely unrealized — series of developments, television seems to be moving further along the anthropotropic road by greatly increasing the access it provides to diverse information. Since in the real world all aspects of the immediately surrounding environment are theoretically perceivable (i.e., there is no hierarchy or exclusivity of information) — and since, in the pre-technological imagination of humans, a myriad of perceptions are theoretically possible — it follows that a fully anthropotropic technology must offer its users a commensurately wide choice of presentations. Yet as anyone who lives in even the "media saturated" area of New York City can testify, the choice of programming offered on television thus far would fail to equal even the most paltry of imaginations. But this limited situation appears to be changing: combinations of new computer and "fiber optic" distribution technology promise, according to some estimates, to deliver access to over 150 television channels presenting a variety of programming on nearly every conceivable subject.[46] Combining even a part of this projection with the home video libraries that are already beginning to develop from the video-recording systems described in the last paragraph, television seems well on its way to providing instant accessibility to a universal bank of information. But as such an accomplishment is still more in the future than the present, a fuller discussion of its feasibility and implications must await Chapter 7 below.

46. Such was suggested by Paul Klein, then former NBC network executive, in a talk about the future of cable television and computers given at The New School for Social Research in New York City in the fall of 1975. Klein envisioned a computer terminal in every home which, when hooked into a coaxial cable system, would put a vast variety of television programming at the consumer's fingertips. (The number of channels in current electromagnetic wave broadcasting is restricted by the number of usable frequencies, whereas with cable, new channels can be created simply by constructing more or bigger cables.) The development of fiber optic technology since Klein's speech, moreover, offers a great increase over both the efficiency and flexibility of cable — and this not only makes

Klein's 150 channel prediction more likely, but opens up the possibility of holographic and interactive transmissions via television, to be discussed more fully immediately below and in the next chapter. See Stewart E. Miller, "Photons in Fiber for Telecommunication," in *Electronics: The Continuing Revolution*, eds. Phillip H. Abelson and Allen L. Hammond (Washington, D.C.: American Association for the Advancement of Science, 197 7), pp. 167-172 for a technical description and explanation of fiber optics; see Maddox, pp. 145-165, for a brief summary of the history and potentials of cable television. For an indication of the availability of computer terminals for the home, see J. S. & A. National Sales Group's *Products That Think* 1977-78 sales catalog, which on p. 16 offers a "Home Library Computer" for $220 plus cartridge accessories. See also Ch. 7, ns. 1 and 2 below.

The increased delivery capacity of fiber optics promises to make television a more complete retriever of the real world in at least two additional ways: (1) Three-dimensional holographic photography, already mentioned above, has thus far been non-transmittable on television — this because the relatively narrow electromagnetic carrier waves (subject in addition to all manner of interference and distance problems) and fuzzy picture resolution of television simply cannot handle the high-intensity information that must be transmitted in order to obtain the holographic image.[47] The heightened carrier energy available through fiber optics, however, would solve some of these problems and thus bring the third dimension much closer to living room screens. (2) Similar problems in image resolution have been cited as impeding public acceptance of Bell's "Picturephone"[48] — and these problems would be similarly alleviated through use of fiber optics rather than conventional telephone wires, which would thus help make telephone with pictures, or interactive video, a household item. Moreover, though it may seem that fiber optics — which transmits information on light waves guided through minuscule glass "cables" — in some ways represents a retreat from the portability of

airwave transmissions to the earlier fixed system of wires, the as yet incipient technology will apparently provide nearly all the flexibility of the airwave broadcast (and, indeed, perhaps more— see further discussion in Chapter 7 below).

47. Stroke, p. 223, explains that full holographic transmission requires a bandwidth (channel) at least four times the magnitude of current broadcast TV channels, and a screen resolution of 700 lines/mm, as compared to the current 525 line system of American and 625 line system of European television. Under such circumstances, the only holograms transmittable are simple, non-moving images. See Chapter 7 below for more.

48. Gregor, pp. 84-85.

* * *

The anthropotropic impetus thus appears to be continuing even — or perhaps especially — in its most fully developed expression to date, television. In describing the emergence of these artificial devices that retrieve more and more of the process and content of real-world communication, the present chapter of necessity has only touched upon the highlights of the anthropotropic saga, therein failing to mention the contribution of the numerous, fascinating, more minor characters. Nothing has been said, for example, of interesting offshoots like the Polaroid, which brought a television-like immediacy to still photography, or of the Xerox, which at last brought the medium of print into the photographic age. Instead, discussion has centered around media like television, photography, and telephone, which were both the first and longest lasting in influence of their "species."

Yet even confined to the seminal rather than derivative media, the tracing of the lineage of anthropotropic development has proved to be a complex task of sorting out relationships in what is, as might be expected, an organic-like interrelated puzzle in which each piece or medium bears some sort of reciprocal connection to virtually all others. Thus the phonograph, as has been seen, served to time-balance the spatial bias of the telephone, audio-balance the visual bias of photography, and in effect eventually join with motion photography to create the more fully replicative medium of talking pictures. Similarly, radio's expansion of the access of interactive telephone actually created a new observational medium, which could then serve as a distributional network for the content of motion photography, thereby engendering television; while telephone has continued nearly unchanged for more than a hundred years, safe in its monopoly on interactive communication. If these developments defy linear, simple cause-and-effect explication, it is because the evolution of extensions of organs and organisms, like the evolution of the organs and organisms themselves, takes place in a multi-faceted, interfacing, simultaneous world.

Yet given these inherent complexities, the foregoing survey has nonetheless managed to trace two great branches of anthropotropic evolution: photography, which, beginning with the still photograph, has successively captured the forms, motion, sounds, and colors — in short, the content — of the real world? and electronics, which, beginning with the telegraph, has increasingly recreated the instantaneous, simultaneous, open-access process of communication in the real world. The first recording, the second distributing; they would eventually combine in the medium of television — which, as suggested above, is still evolving towards closer consonance to the real world, with photographic holography offering more realistic content, and fiber optics more realistic distribution.

But a theory must do more than describe developments —

it must discern in these developments certain mechanisms, principles, which may be abstracted from their initial contexts and fruitfully applied to explain and perhaps predict other developments. Several such principles have already been articulated both in the review of media theories in the earlier chapters, and the discussion of primitive "Stage B" technologies in the present chapter. Thus, the distorting impact of some media upon human societies emphasized by media determinists, for example, was explained by a "principle" of media evolution which suggests that the more primitive the medium, the more distorting its effects (end of Chapter 2 above); while the triumph of phonetic over hieroglyphic writing was seen as evidence of a striving in primitive technology for an increase in extensional power at the expense of all other communication aspects (and thus identified, the principle was then useful in explaining the reduction in reality-content in the change from print to telegraph, see see beginning of the current chapter).

Beneath the ensuing parade of anthropotropic media just observed here, other such principles have been operating — determining, if not the route, who marched in the parade and for how long — why, for example, the primordial still photograph and telegraph have survived while later media such as silent movies have not, why radio has not only survived the advent of television but remains relatively intact while television changes, why radio will probably continue to survive in its present form while telephone profoundly changes, and, not to forget the origins of all this, why the phonetic writing on this page remains — and perhaps will remain — the primary medium for intellectual discourse.

Through scrutiny of selected examples such as these, the following chapter will attempt to more fully uncover and articulate the principles that govern the anthropotropic

evolution of media.

CHAPTER 6
PRINCIPLES OF MEDIA EVOLUTION: SURVIVAL OF THE CLOSEST FIT

MEDIA IN HUMAN ECOLOGICAL NICHES

The Roaring Twenties in America were the heyday of what might be termed "one-dimensional" or "uni-sensory" media: at this juncture, two media dominated the world of observational entertainment — silent movies and radio — and complemented each other in that one dealt primarily in pictures, and the other exclusively in sound. Within a few years, two new media appeared on the scene — talkies and television — and in integrating sight and sound in coordinated packages, broke forever the hegemony of silent movies and radio. Yet as the dust still clears on this slow revolution, there comes to light a fascinating observation: namely, that while silent movies exist now only in the after-life of art and nostalgia, blind radio has somehow managed not only to survive but thrive in the face of sighted television.

How can an anthropotropic theory of media evolution explain this development? In effect, the survival of radio and demise of silent movies means that hearing without seeing is somehow a more apt media mode than seeing without hearing. And since the anthropotropic perspective suggests that media evolve towards increasing replication of the pre-technological world, it follows that the aptness or success of hearing without seeing in media must be tied in some way to the pre-technological world — that sound without sight in media in some way reflects some essential element of real-life communication. And indeed, an examination of the pre-technological environment turns up just such a pattern of communication.

In the pre-technological world, hearing without seeing is both physiologically convenient and common in communication. To hear without seeing, we have but to close our eyes; we avail ourselves of this mode every time a sound wakens us from sleep — and, in effect, when we listen to stories in the darkness, eavesdrop on the next room, cock our ears to hear what's going on over the next hill, and so forth. In contrast, to see without hearing is physiologically awkward and all but unheard of in the natural environment: sight is almost always accompanied by some concurrent sound in the real world, and to even attempt to block out sound while maintaining sight we must resort to the uncomfortable, inefficient posture of sticking our fingers in our ears. Hearing without seeing thus seems a part of the pre-technological repertoire of communication, while seeing without hearing does not.[1] (Moreover, the primacy of hearing would have great survival value in the pre-technological world, which grows dark every night but never really silent.)

Applying this observation to the problem of radio's success and silent movies' failure, an explanation immediately suggests itself: to wit, that radio survives because it approximates a mode of pre-technological communication, while the silent movie died because it did not.

1. George Gordon makes essentially the same observation in his distinction between the "narrative-first" and "picture-second" mediums, writing in *Persuasion* (New York: Hastings House, 1971), p. 13 that "Visual experiences may be employed to enhance (or even create) narratives, but the pictoral element may be severed from *any* communication. All we have to do is shut our eyes." (Italics in original.)

And this explanation, in turn, suggests a general principle of media evolution: that a medium's survival quotient is directly proportional to its approximation of the pre-technological, human communication environment. While all media evolve towards increasing replication of the real world, some media — perhaps even early anthropotropic specimens — manage to attain a close consonance with some aspect of the real communication environment, as radio did to the pre-technological pattern of hearing without seeing. These media then survive relatively unchanged, while contemporaries and even successors continue to undergo more drastic transformations in evolution. This situation, as might be expected, is analogous to biological evolution, in which ancient organisms like the star-fish exist unchanged for millions of years, while other organisms evolve either into extinction or out of recognition. The surviving organisms are said to have attained an "ecological niche" — a satisfactory response to the requirements of a particular environment — while the others have yet to "find" theirs. In the evolution of media, as suggested in Chapter 3 above, the environment to which media must respond is the pre-technological world, or the human preference or desire to retrieve the patterns of real-world communication; a medium may therefore be said to have found its "human" ecological niche when it has achieved a satisfactory response or approximation of some essential aspect of the pre-technological world.

Such an "ecological niche" principle of media evolution, thus derived from the case of radio, should and can explain the unlikely survival — or demise — of other media. The development of photography provides an example in need of just such explication. As described in the previous chapter, the original photograph, both motionless and black-and-white, was eventually succeeded by photographic images that moved and had color. In the aftermath of these more fully replicative descendants, however, one of the original forms of photography survived while the other withered: thus, still photography flourishes in the form of snapshots alongside of motion photography and television, while black-and-white has all but totally faded (except as art/nostalgia) in the commercial glow of color pictures. Why did still survive motion but black-and-white not survive color? If, as the ecological niche principle suggests, media survive when they achieve correspondence to some pre-technological communication pattern, then the answer must lie in the pre-technological environment, where stillness is indeed a common perception and lack of color is not. Thus, the world seems motionless whenever we gaze into the far distance or simply look at a close-by object or scene at rest, but black-and-white in the real world rarely intrudes upon color except in some special instances of twilight and shadows. Still photography, then, has captured a fundamental aspect of communication in the real world, whereas black-and-white photography simply has not — and thus still photography thrives while replication in black-and-white languishes.

Yet another example of media attaining human ecological niches can be found in the recent evolution of radio and the replication of sound, which, as noted in the previous chapter, now provides a multi-dimensional blanket of sound that surrounds the listener by approaching from numerous angles.[2] On radio, this multi-dimensional sound is best transmitted on FM channels, which are static-free and hence sensitive to nuances in sound, and allow for at least a stereophonic and in some cases quadraphonic (four-way) speaker-system to disperse the sound. Yet given these obvious anthropotropic advantages of FM stereophonic radio, AM monaural (one sound perspective) radio has nonetheless managed to survive quite well.[3] Why?

2. See Zilkha, p. 67, for a brief description of recent advances in audio presentation.

3. Sol Taischoff, ed. *Broadcasting Yearbook 1977* (Washington, D.C.: Lawrence B. Taischoff, 1977), p. A7 lists 4,463 AM stations and 3,571 FM stations (including 804 non-commercial FM stations) licensed in the U.S. in 1976.

Aside from the still important extensional benefits of AM radio — which can be transmitted much further than FM, and is especially effective at night — the answer should at least in part again be found in the pre-technological world. While sound in the real world does indeed in most cases approach from a 360 degree horizon or a multiple of perspectives, there does seem to be at least one significant exception in which sound in the pre-technological environment tends to approach the listener from one, monaural-like source: namely, the sound that emanates from the human mouth. In attending to speech, unlike the general sound environment, the listener tends to focus on a specific, readily identifiable speaker or sound source — thus in effect singling out a monaural source from the general environment. Has AM monaural radio survived because it replicates this pinpoint talk perspective?

Recent programming alignments in radio suggest that such may be the case. Since the commercial surge of FM radio in the 1960s (sparked by the FCC's ruling in 1965 that FM must offer programming substantially different from AM),[4] AM stations have tended to adopt all-news, listener phone-in, or other talk formats; while FM stations, in replication of the more anonymous and multi-perspective blend of sounds in non-speech pre-technological "soundscapes," have usually adopted some sort of musical format (indeed, NBC's recent attempt at an all-news FM radio network was a dismal failure).[5] Thus, AM and FM survive side by side because each has found an acceptable ecological niche: AM monaural in the replication of talk environments, FM stereo in the replication of general sound environments through music. [6]

4. See Sidney W. Head, *Broadcasting in America*, 3rd ed. (Boston: Houghton Mifflin, 1976), p. 150.

5. In the New York City area, for example, at least two AM stations (WINS and WMCA) have changed from music to talk since 1965, and no AM stations have changed from talk to music; while at least one FM station (WNWS, the NBC network affiliate) has failed in a talk format, and another (WPLJ) changed its midnight to morning shift from talk to music during the past few years. Nationally, Taischoff, p. D 78, lists 85 AM and 15 FM stations licensed for all-news in 1976, with an increase of 10 AM and 2 FM all-news stations in 1977 (this from Taischoff's 1978 *Yearbook*). As for other types of talk formats (such as interviews, phone-ins, etc.), Taischoff, p. D 81 reports 186 AM and 102 FM licenses; but the usefulness of these figures is questionable, since Taischoff defines these non-news talk formats merely as those that feature "20 or more hours of talk per week." This makes it difficult to determine whether a station with 25 hours a week of talk and more than 100 hours a week of music has survived because or in spite of its talk segment. For a brief report on the failure of NBC's all-news, primarily FM, network, see "NBC to Discontinue Next Year Its All-News Service for Radio," *New York Times*, 4 November 1976, p. 78.

6. This is not to suggest, however, that the public has an appetite for more than a limited number of talk-format stations, or that the public's desire for replication of talk environments via radio is necessarily even a continuing one. Indeed, the public's general preference for music over talk — music in any form — no doubt explains the continued preponderance of music formats on AM (though, as has been seen above, this preponderance is proportionately less than the preponderance of music formats on FM). How, then, might the continued playing of music on AM monaural radio affect the evolution and future content of radio? One possibility would be for AM to gravitate to a coarser, more "uni-

dimensional" type of music — and indeed, the tendency of AM to play "Top 40" music, and FM to play classical, jazz, and more sophisticated types of rock music suggests that such an alignment is already underway. Moreover, the recently raised possibility of AM becoming stereo and FM becoming quadraphonic (see "In the FCC's Lap: FM Quad, AM Stereo, TV Stereo," *Broadcast Management/Engineering*, October 1977, p. 53) suggests a further accommodation of radio to subtle varieties of music and non-talk sound environments. Of course, if AM stations do go stereo, the continuation of talk formats on these stations would be as much of an overkill, anthropotropically speaking, as stereophonic telephones.

Thus the evidence strongly supports the proposition that in the evolution of media — as in the evolution of organisms — a medium will survive, regardless of what comes after it, if the medium has achieved a close correspondence with some aspect or pattern of pre-technological communication.

PRECISION VS. SCOPE OF APPROXIMATION

Yet the question remains as to what constitutes a "close" or "satisfactory" approximation or correspondence. In each of the three cases cited above (that of radio in general, still photography, and AM and FM radio in particular) the medium survived relatively unchanged by replicating a rather narrow range or element of the pre-technological environment (in these three instances, eavesdropping on sounds, stillness of images, and talk and general acoustic landscapes, respectively). And yet, at the same time, media that attempted to replicate a much broader part of the pre-technological environment — such as the coordinated sights and sounds of motion pictures and television — have continued, as described in the preceding chapter, to undergo a marked, form-changing evolution.

Thus, the television image has increased to ten times its recent size (see above), and motion photography reaches out for the holographic third dimension (see above). This suggests that the more of the real world a medium attempts to replicate, the more it must replicate in order to survive unchanged.. Having captured motion and sound and immediacy through television, the human appetite is only further whetted for an all-encompassing medium that retrieves all aspects of pre-technological reality — and this provoked desire fuels the continuing evolution of television.

Meanwhile, still photography and radio, having not even

a whisper of retrieving the full pre-technological world, endure in a comfortable corner of the media galaxy. Thus, to the general principle that media survive in proportion to their approximation of the pre-technological environment, an important corollary must be added: namely, that the precision with which a medium replicates even a small segment of the pre-technological environment is more germane to the medium's survival than the scope with which the medium attempts to replicate the pre- technological world. In other words, the "close" correspondence to the real world that a medium needs to survive entails accuracy and not necessarily range.

THE PRINCIPLE OF "NET GAIN"

The requirements of a successful ecological niche for media may be further defined by examining the seemingly strange case of 3-D movies, already alluded to in Chapter 1. As suggested, the virtual demise of 3-D movies shortly after its inception in the 1950s seems to contradict the proposition that media survive in proportion to their correspondence to the real world — since in providing the third dimension of depth, 3-D obviously replicated at least one aspect of the real world not found in other flat-screen, two-dimensional media.

It turns out, however, that to provide this heightened replication of reality in one area, 3-D movies had to reduce the replication of reality in other areas: thus, to perceive the 3-D effect, a viewer must don special, ungainly glasses, and keep the head in a relatively fixed, immobile position (neither of which are required in the real world, or in the viewing of two-dimensional media). The consequent failure of 3-D movies, then, suggests that for a medium to survive, it must not only achieve a close approximation of some pre-technological pattern of communication, but achieve a close approximation that is a net gain over the approximations attained by previous or concurrent media. Media that attain consonance to reality in one area by distorting reality in others are not likely candidates for survival, and indeed may be considered "throw-backs" of a sort to the pattern of evolution in primitive, "Stage B" technologies, in which media gain in extensional ability through the sacrifice of the real world. (The survival of radio represents a partial exception to this "net gain" principle, in that radio, as indicated in Chapter 5, provided mobility in communication at the expense of the interactive capacity already present in the telephone. An explanation for radio's survival in these circumstances may lie in the sheer magnitude of its mobile advantages, in effect more than compensating for the loss in interactive ability, and thus still constituting some sort of "net gain" in retrieval of reality.

It is worth noting again, however, that the loss of interactive capacity in radio has been the source of much evolutionary pressure on radio to regain interactive communication, as currently evidenced by "C.B." radio, "phone-in" radio formats, etc.) In the technological retrieval of the third dimension, application of this "net gain" corollary to the ecological niche principle suggests that holography, which unlike 3-D movies offers the third dimension of depth to the naked eye from all perspectives, may be a much more viable medium than 3-D. (See Chapter 7 below for further discussion of holography.)

REPLICATION VS. EXTENSION

It is important to bear in mind that while the main evolutionary thrust of anthropotropic media is an increasing retrieval of the real world, they, nonetheless, do so while attempting to maintain (or even heighten) the extension across time and space achieved by more primitive media. Just how this extensional factor figures in the survival of media is suggested in the example of the telegraph and telephone. As indicated in Chapter 5, the telephone in all ways represented a net anthropotropic gain over the telegraph — maintaining the instantaneous distribution system of the telegraph, while improving on its content with the substitution of human voices for dots and dashes. Why then has the telegraph continued to survive?

Perhaps because, until recently, the telegraph provided a permanent record of its transaction unavailable through telephone. In other words, the telegraph, for all its distributional or spatial capacities, maintained an extension across time — a permanency — that telephone lost. This suggests that media will survive, even if they offer a lower level of replication than subsequent media, if they provide an extension across time or space that subsequent media fail to maintain. This lesson is highly important, because it emphasizes that the purpose of anthropotropic evolution is not merely a retrieval of the pre-technological environment — or a return to the reality of biological constraints upon communication — but an attempt to extend the pre-technological world (rather than substitute for it, as in primitive technologies) well beyond the original biological boundaries of that world. This, in effect, is the source of the distinction in the present work between "Stage A" or pre-technological environments, and the environments created by "Stage C" or advanced technologies. The continuing extensional capacity of "Stage C" anthropotropic technology also underscores the profoundly conservative nature of these media — seeking not only to retrieve or conserve the elements of pre-technological communication, but to maintain or conserve the extension across pre-technological boundaries which was the initial (and continuing) purpose of technologies.

Thus, a medium that retrieves an aspect of reality while failing to maintain extensional capacity falls short of the ecological niche. In the case of telephone, its outlook for survival has been enhanced with the recent addition of time-binding, tape-recording technologies (e.g., home answering systems, etc.). This development has of course at the same time pulled the rug out from under the telegraph, by depriving it of its last advantage — extensional or otherwise — over the telephone. And indeed, the effects of this coup de grace are already being felt in the telegraph industry, with Western Union reporting income from telegrams down to a new low of 11% of total revenues m 1976.[7]

7. "A Space-Age Treadmill," *Forbes*, 1 August 1977, pp. 59-60. See, however, the end of the current chapter for evidence that phonetic writing may have attained its own ecological niche. To the extent that such is the case, the written communication of print and telegrams may yet have some survival possibilities.

CO-EVOLUTION AND CONVERGENCE

Thus the survival potential of any medium, much like the evolution of media in general, is a complex affair dependent not only upon the medium's ability to replicate reality and extend across space and time, but upon how the medium's performance in these areas compares with the performance of other media in similar areas. And this in turn raises the question of whether media may attain an ecological niche synergistically — that is, whether two or more media, imperfect replicators of reality individually, may in concert manage to closely approximate some pre-technological pattern. The survival of telephone and radio — each, as has been seen, an imperfect replicator on its own — may be due to what has been, in effect, such cooperation, with radio supplementing the access limitations of the telephone, and telephone making up for the interactive deficiencies of radio. Other examples of such cooperation have already been noted, such as the space and time balancing act of the telegraph and photograph, the ear and eye act of the phonograph and photograph, the space and time complement, again, of telephone and phonograph; and more recently, the space and time interaction of radio and phonograph records (and audio tapes), and television and home video systems (including video-discs). In all these instances, the partners achieve a greater consonance to some human communication pattern than either partner can achieve on its own.

An interesting pattern, however, suggests itself when the development of these cooperative media is traced. Whereas in the earlier examples, such as telegraph and photograph, radio and telephone, and so forth, the partners complement each other at a distance (i.e., they function independently of each other), in the more recent examples such as radio and phonograph records or television and home video, the media cooperation takes place as part of a unified, integrated, single process. Thus, phonograph records (and audio tapes) provide the actual programming content for radio, and home video recorders are intended to be attached to the actual TV system. These examples suggest that it may be the nature of cooperating media — perhaps indeed of media in general — to increasingly converge into single, more complex technologies that increasingly replicate more of the real world as unified, multi-faceted systems. Such a convergence principle of evolution would certainly explain the development of television itself — which, as indicated in the preceding chapter, is in effect a convergence of the content of film with the distribution of radio — and would be consistent both with Teilhard's view of communication technologies as forming one living "superbrain" around the earth, and with the metaphor of computer as brain, and television cameras and microphones as eyes and ears, etc., of one integrated, life-like media system (see Chapter 4, above). [8]

8. A seemingly opposing view appears in Gary Gumpert's "The Rise of the Mini Comm," *Journal of Communication* 20 (September 1970): 280-290, which suggests that as media develop they present increasingly specialized programming to increasingly specialized audiences. Examples of such fragmentation would be the shift in music programming on radio (largely since the activation of FM radio in the 1960s as described earlier in this chapter) from a general, "popular" music or "Top 40" format to highly differentiated formats such as

"progressive" rock, "easy" rock, jazz-rock, and so forth; and the increase in varieties of television programs made possible by new cable systems, briefly mentioned in Chapter 5 above. Such specialization in usage and content, however, by no means indicates a specialization in the medium's structure or its capability for processing information; indeed, quite the reverse seems true— that is, media which provide the most individualized services are often those with the most integrated, or generalized, structures. Thus, fine shades of music are best presented on static-free FM radio, which in effect combines the distribution of AM radio with the clarity (lack of static) of the phonograph; and similarly, cable TV combines the audio-visual environment first produced by broadcast TV with the access possibilities of wires first used for telephone (see Chapter 7 , "Mobility and Access," for more on cable, and fiber optic, TV distribution) . The evolution of media towards more integrated systems thus seems to enhance, and even make possible their capacity for more personalized programming noted by Gumpert. Moreover, since all pre-technological communication is personalized or personal (though not private), the growing ability of media for more personalized services is entirely consistent with, and expected by, anthropotropic theory. (See also chapter 4 above for a comparison of media convergence with Popper's views on the evolution of knowledge towards more integrated forms.)

* * *

Thus, the examination of various examples of media survival and demise in this chapter has suggested the following series of interrelated principles that seem to govern the evolution of media in general: (1) that a medium's survival or attainment of a "human" ecological niche depends upon the medium's ability to closely approximate some pre-technological pattern of communication (derived from the examples of radio vs. silent movies, still vs. black-and-white photography, and AM vs. FM radio); (2) that the precision with which a medium approximates some aspect of the pre-technological environment is more crucial to the medium's survival than the scope of approximation (from the example of radio vs. television); (3) that for a medium to survive, its approximation of reality must represent a net gain over the approximations of prior media — that is, a medium that heightens approximation of pre-technological environments in one area but reduces the approximation in other areas is not a good candidate for survival (example of failure of 3-D movies); (4) that a medium's survival depends not only upon its ability to retrieve reality, but its capacity to maintain the extensions of previous media across time and space (example of telegraph vs. telephone); (5) (a) that media may survive by approximating the pre-technological environment in concert, and (b) media seem to be evolving towards increasing convergence in which single media systems perform a variety of extensional and replicational tasks (early examples of telegraph and photography, etc.; recent examples of television, computers, etc.). (Note that the consequence of principles 5b and 2 suggests that the most advanced media — i.e., those which like television exhibit the greatest amount of convergence — are also most likely to continue drastic evolution since they tend to replicate larger and larger configurations of pre-technological communication, without necessarily attaining a sufficient degree of precision.)

When these primarily "Stage C" or anthropotropic principles are combined with some of the observations made earlier in connection with more primitive technology — e.g., that primitive technologies evolve towards time and space extension in total disregard of replication of reality; or that the more primitive the technology, the more distorting it will be to its human users (the second principle obviously follows from the first) — a complete picture of the workings of technology from its origins to the present, with strong implications for the future, begins to emerge.

PERSISTENCE OF ABSTRACTION

Before attempting to apply these principles towards predicting the future development of media, however, it might be useful to see how they might explain a particularly troubling example of media survival, one that reaches back past even the origins of physical technology itself. The problem arises from the 5,000 year plus survival of the highly abstract technology of phonetic writing: how, if media survival is keyed to replication of the pre-technological environment, could such a poor, indeed almost non, replicative technology as phonetics have survived?

The dimensions of this contradiction may be better appreciated by recalling McLuhan's observation (beginning of Chapter 4 above) that phonetic writing is not one, but two levels of abstraction removed from reality: thus, the written word "tree" bears no discernible connection to the spoken word "tree," which in itself has nothing in common with the tree that grows in the pre-technological backyard. And yet both speech and phonetic writing survive, while technologies like silent movies and black-and-white photography, which provide much closer if far from complete approximations of the tree in the pre-technological world, have all but disappeared.

It is almost as if a whole phylum of non-replicative technologies — beginning with speech, further abstracted in phonetic writing, and further extended (though not further abstracted) through print — flourish in exemption of the principles of media evolution discerned in the present chapter. (And note that the survival of these media is not a case of more primitive technologies having extensional advantages over later technologies — as in the permanency of the telegram as compared to the telephone call — since the printed silent photograph has the same amount of extensional power as the printed word, and far more than speech and unprinted phonetic writing.)

The view of a branch of abstract technologies as exempt from the general evolution of media suggests at least one way of dealing with the problem: simply sever the exempt branch from the general body of the theory. Thus, the principles of media evolution described in the present study could be said to apply only to the electronic and photo-chemical branches of media first introduced in the early part of the 19th century — anthropotropic theory thus becoming a theory only of replicative, or replicative-intended, media, with all other technologies subject to some other frame of explanations. A less extreme and more plausible pruning approach might insist, alternatively, that all technologies are indeed subject to anthropotropic evolution — and define at least speech and perhaps phonetic writing out of the problem by arguing that they are not in any tangible sense technologies. Such an approach would ironically suit those theorists who, like Mumford, draw a sharp distinction between language and tools, viewing language as the more fundamental expression of humanity; and such a distinction at the same time would allow anthropotropic theory the dignity of claiming to pertain to all communication technologies, rather than a selected few. [9]

9. "'Long before he had achieved the power to transform the natural environment, man had created a miniature environment, the symbolic

field of play," Mumford writes in *The Myth of the Machine, vol. 1: Technics and Human Development* (New York: Harcourt Brace Jovanovich, 1966), p. 8. Mumford continues, "the evolution of language — a culmination of man's more elementary forms of expressing and transmitting meaning — was incomparably more important to further human development than the chipping of a mountain of hand-axes. Besides the relatively simple coordinations required for tool-using, the delicate interplay of the many organs needed for the creation of articulate speech was a far more striking advance." Moreover, Mumford, as seen on p. 219 of the present study, apparently views writing and print as expressions of language, while he in effect places the replicative "electronic" media in the inferior category of tools, both less important to human development and capable of greater misuse than language. Such a perspective would tend to make anthropotropic theory consistent but somewhat trivial — consistent in that it does indeed pertain to all communication technologies (or tools, as opposed to language), but trivial in that technologies would be seen as less significant communicators than language. It is perhaps Mumford's very view of technologies as second-class communicators that accounts for the lack of attention he has given his own observation of anthropotropic evolution of media made more than 40 years ago (see end of ch. 3 above). (See also ch. 4, n. 12 above for an opposing view of communication technologies as languages.)

Justification for perhaps a more plausible pruning of this sort has come from Walter Ong, a former student of McLuhan's, who argues that while writing and print are technologies, speech most certainly is not, because

> the real word, the spoken word, in a profound sense is of itself bound to ongoing, lived human existence. . . . The spoken word, however abstract its signification or however static the object it may represent, is of its very nature a sound, tied to the movement of life itself in the flow of time. . . . Languages, in which words originate, . . . require no external technological

> skills at all. They are not 'out there,' distanced. Languages come from within and they are distinctively human in that, among other things, they require man's own kind of oral and vocal apparatus.[10]

On the other hand, "writing and print are technologies, requiring reflectively prepared materials and tools. . . . they have sequestered the essentially participatory word — fruitfully enough, beyond a doubt — from its natural habitat, sound, and assimilated it to a mark on a surface, where a real word cannot exist at all."[11] Such a distinction has in effect guided the development of anthropotropic theory thus far: speech has been viewed as a pre-technological (although abstract) pattern of communication, first extended and distorted by the primitive technology of phonetic writing, then retrieved (and further extended) by the advanced technologies of telephone, radio, etc.(see, for example, Chapter 5 above). Categorizing speech as pre-technological, moreover, both appeals to common sense, and seems to explain the success of an abstract, non-replicative means of communication in a world otherwise dominated by media that increasingly replicate the pre-technological environment.

10. *Interfaces of the Word* (Ithaca, N.Y.: Cornell Univ. Press, 1977), pp. 20, 22.

11. Ibid., pp. 22, 21.

Non-replicative speech, in other words, survives in a world of replicative media because it happens to be part of the original environment being replicated. Such a solution is possible, of course, only with Ong's definition of speech as non technological even though it is abstract, or, to use Ong's words, as "real" (or "natural") and abstract at the same time. But this definition tacitly assumed thus far in the present study, presents several serious problems. To begin with, Ong contends that the spoken word is "real" because it is "a sound," and "distinctively human" because it is a sound created by human organs (as opposed to human tools or products); more importantly, that this "reality" and "humanity" assert themselves in speech "however abstract its signification." According to Ong, in other words, speech is fundamentally human not because of, but in spite of, its abstraction. Yet surely the quality that most distinguishes speech as a communication system is its ability to describe situations to which it has no "real" or "natural" connection (as opposed, for example, to shrieks on the one hand, and photography on the other). Thus Ong, in effect, makes a strong case for the "reality" and "humanity" of grunts and any other sounds produced by human organs, with speech included only on the basis of its most trivial characteristics. But even a distinction between speech as human-produced and hence human (non- technological), and writing as tool-produced and hence technological, seems at the core rather flimsy: speech is produced by the action of human vocal chords upon the atmosphere; writing may be produced by the action of human fingers upon wet sand (but usually by the indirect action of human fingers guiding a tool upon some other surface); what, in terms of origin and even execution, is the effective difference between these two activities? To be sure, speech is fleeting and writing is not. But evanescence seems no more "tied to the movement of life itself" (as Ong suggests above) than permanence: the real environment, from the human perspective, obviously has elements of both permanency and change. All considered, then, McLuhan's insistence that all human artifacts — whether spoken, written, or electronically produced, whether created directly or indirectly by human organs — are extensions of human systems and hence technologies, seems less of a distortion than Ong's withholding of speech from the technological realm. [12]

12. McLuhan's view that "all of man's artifacts, of language, of laws, of ideas and hypotheses, of tools, of clothing and computers — all of these are extensions of the physical human body [in which] . . . the mind of man is structurally inherent,"("Laws of the Media," pp. 175-176) is shared in many ways by Popper's conception of "World 3," "the world of the products of the human mind" consisting of "language, theories, works of art and science (including technology)," etc. (*The Self and Its Brain*, pp. 15-16; see also ch. 4 of the present study). Note also that, unlike Mumford and Ong, neither McLuhan nor Popper view technology as in any fundamental way non-human. Indeed, both emphasize the extent to which humanity is the consequence of human products, though McLuhan, unlike Popper, often depicts humans as the passive partner in this relationship (see chs. 2 and 4 above).

Thus, the attempt to account for speech's persistence by making the seemingly reasonable assertion that speech itself is pre-technological fails, because upon closer examination speech seems to have more in common with technologies than not. And, in any case, such a solution would have addressed only the survival of speech, offering no explanation as to why phonetic writing, a primitive distortion of speech, has survived the advent of technologies such as telephone and radio which provide a far more accurate replication of the spoken word. If the principles of media evolution discerned in this chapter are at all valid, then the survival of phonetic writing must lie in its replication of some continuing aspect of the pre-technological environment other than speech — an aspect that telephone and radio do not as closely replicate. Moreover, once speech is (correctly) recognized as technological, the explanation of its survival must also lie in its replication of some continuing aspect of the pre-technological environment (obviously something other than speech itself). The possibility suggests itself that speech and phonetic writing, appearing relatively near each other in media history, may both be replicating the same element of the pre-technological environment (phonetic writing visually balancing the acoustic extension of speech); and since the characteristic that speech and phonetic writing have most in common is their high levels of abstraction, the further possibility suggests itself that their common human ecological niche may be the process of abstraction itself.

Thus far in this study, the pre- or non-technological communication environment has been defined by an almost randomly chosen series of examples, rather than an exhaustive inventory;[13] such a procedure reflects both the exploratory nature of this study, which gained a better understanding of the pre-technological environment as the study developed, and the inevitable degree of ambiguity in a construct such as "pre-technological environment." Nonetheless, several dependable "classes" of pre-technological constituents, which overlap in many respects, have been repeatedly identified here: these include pre-technological "contents," such as color, the third dimension, sight-and-sound, and so forth, retrieved for the most part by the photographic branch of media; pre-technological "processes," such as simultaneity, interaction, and immediacy, retrieved primarily by electronic media; and the most obscure constituent, imagination, which, as suggested earlier in this study, serves not as a source for literal retrieval or replication by technologies but as a goad for the technological process of extension itself. Might there in addition be a pre-technological process of abstraction, a deliberate divergence from literal patterns of reality that speech and phonetic writing (and perhaps other systems such as art) accurately reflect?[14] Can the survival of speech and phonetic writing, in other words, be explained as the survival of primitive technological processes which reflect — regardless of their content — a pre-technological or human process that might be termed "abstract thought," in much the same way as telephone has been said to survive by replicating the pre-technological process of interaction?

13. Developing a theory without the prior benefit of precise and exhaustive definition of terms may seem contrary to standard scientific procedure and even logical argument, but is defended by Popper and Eccles, who write in *The Self and Its Brain*, p. viii , that "what we are interested in is not the meaning of terms but the truth of theories; and this truth is largely independent of the terminology used." But how can a theory that seeks to explore the relationship between two variables — in this case, the pre-technological environment and technological communication — hope to discover and describe the "true" relationship when the variables themselves are incompletely defined, as evidenced in the present problem of how to classify speech and phonetic writing? The difficulty can be overcome, as the present study itself perhaps demonstrates, if there are at the start at least some certain, representative examples of each variable; and once the relationship between these obvious examples has been determined, it can help clarify the status and connection of the more ambiguous cases. Thus, the present study began with the unassailable definition of color and depth perception as pre-technological, and the equally unassailable definition of film and TV as technologies; having discerned how these and other undeniable members of the second group have evolved in response to undeniable members of the first group, the study can now "work backwards" and better explain how less certain examples such as speech fit into (and perhaps modify) the anthropotropic equation.

14. Note again that imagination, as viewed here, is "unreal" only in its transcending of time and space (which serves as the model for technological extension), and not in its portrayal of other elements of real communication environments. For example, we might imagine — or "picture" — ourselves standing beneath a tree in a city across the ocean, and most likely we would see ourselves and the tree in three-dimensions, real-life colors, and so forth (unless we had some conscious or unconscious purpose in distorting — i.e., abstracting or symbolizing — the images). On the other hand, to think abstractly about such a

situation — as opposed to merely imagining it — might be to view it as a mathematical distance, or an "escape" from unhappiness; in short, in representations or symbols that have no literal or necessary connection with the physical image of ourselves under the tree (precisely the arbitrary relationship that speech and writing have to the real objects that speech and writing represent). Of course, most mental processes involve a co-mingling of imagination and abstraction — and in many, no doubt, their contributions are hopelessly intertwined — but it nevertheless is helpful, in seeking the antecedents of technology, to attempt to so distinguish between the two faculties. (As to the relative input of the two in human activities, the present study suggests that technology is becoming more imaginative — i.e., more responsive to imagination — and less abstract, while modern art consequently grows less imaginative and more abstract. For more on technology and art, see ch. 1, n. 30 , and ch. 4, n. 49 above; see also the latter note for early 19th century philosopher Abercrombie's description of imagination as a mental rearrangement of "real scenes.")

To qualify as "pre-technological" communication in the present study, prospective candidates have thus far been obliged, in effect, to meet two requirements: (1) does the communication pattern in question exist, at present, independently or in the absence of the technology (or technologies) said to replicate it; (2) did the communication pattern exist, historically, before the development of the technology said to replicate it. Color perception, for example, has been regarded as pre- technological because humans obviously see in colors in the absence of photography and television, and just as demonstrably saw in colors before these media were invented (we know this from written references to color, etc.); similarly, interactive communication can take place now without the telephone, and undeniably took place before the telephone. In the case of abstract thought as a pre-technological determinant of speech and phonetic writing, however, the candidate passes only the first test cleanly: although abstract thought

is clearly capable of occurring without actual speech or writing at present, the practice of abstract thinking before the advent of speech in our remote history is far from demonstrable.

What would constitute evidence of the existence of abstract thought before the development of speech? "Cognitive" anthropologist Alexander Marshack has examined 30,000 year-old Cro-Magnon cave paintings and sculpturings and concluded that they performed a symbolic, perhaps religious function (i.e., they were intended and used as more than literal replications of their subjects); apparently even more abstract and ancient are the zig-zag or "repeating v" patterns that have been found not only on Cro-Magnon artifacts, but on a piece of ox rib 300,000 years old, and which Marshack believes were some sort of recording or counting system (such a counting system would be highly abstract in that the "v" marks bear no physical resemblance to the sunsets, animals, or whatever being counted).[15] An equally early origin is suggested by another line of evidence explored by Glynn Isaac, who in a recent *Scientific American* article describes a shift "over the past several hundred thousand years" from "opportunistic" or natural to more "arbitrary" tool forms — that is, a change from tools that were apparently little more than raw stones chosen from the environment due to their convenient form, to tools that were fashioned from stones in increasingly standardized and apparently culturally-dictated forms.[16]

15. For a brief description both of Marshack's work and a display of Cro-Magnon artifacts at the American Museum of Natural History (on exhibit from May 1978 to January 1979), see Boyce Rensberger, "The World's Oldest Works of Art," *The New York Times Magazine*, 21 May 1978, pp. 26-29, 32, 37, 40, 42. Marshack's main work in this area is *The Roots of Civilization* (New York: McGraw-Hill, 1972).

16. " The Food-sharing Behavior of Protohuman Hominids," *Scientific*

American 238 (April 1978): 104. However, although Isaac indicates that the later tools were humanly constructed and hence artificial in creation, he fails to specify whether the shapes of such tools were abstract, or merely imitations of the most useful of the earlier, natural tool forms. To the extent that the later tools were attempts to standardize the best of the forms already present in the natural environment, the construction of these later tools would seem evidence more of a replicative than an abstract mentality, and hence would be of little assistance to the present consideration of the origins of abstract mentality. See also note 19 below in this chapter.

The beginnings of abstract thinking thus appear to reach well back in human history; but does this mean that they pre-date spoken language? In the judgement of the above anthropologists, at least, just the reverse seems true: for the existence of artifacts that reflect abstract thinking is believed to presuppose, *a fortiori*, the existence of language. A recent review of Marshack's work with Cro-Magnon creations thus concludes that "in order to conceive and use such objects, Cro-Magnon peoples must have had a complex language that was capable not only of expressing great varieties of observed detail but also of grouping the details into broad, organizing concepts";[17] and Isaac suggests "that the principal evolutionary change in the hominid line leading to full humanity over the past two million years has been the great expansion of language and communication abilities, together with the cognitive and cultural capabilities integrally related to language."[18] In effect, then, the evidence cited above primarily demonstrates the venerability of speech itself, and offers no clue as to whether abstract thinking existed before the development of speech. [19]

17. Rensberger, p. 40. Marshack's own view (*The Roots of Civilization*, pp. 117-118) is a bit more complex, but may be excerpted and summarized as follows: (a) human (and "near-human") communication involves not only

the "ability to vocalize sounds that are recognized as symbols with meaning," but a "non-verbal process" that includes "mimetic and kinesthetic understanding" (i.e., non-abstract) as well as a capacity for "synthesizing and abstracting relational concepts"; (b) the non-verbal cognitive/communication process may have been operating even before humans (or near-humans) achieved the physiological ability to speak; but the development of speech, in any case, was "only one aspect, perhaps simultaneous, of that broad, evolving, non-verbal process involved in communication and symbol-making"; (c) the paintings and other cognitive artifacts examined by Marshack (he calls these "art and notation") are products of that "broad, evolving" communication process that includes speech. Since only the last of these conjectures is an interpretation of physical evidence, and since the specific interpretation that Marshack gives to this evidence offers no guidance as to whether abstract thinking preceded speech (both are regarded as prerequisites of the artifacts), Marshack's suggestion that some type of non-verbal abstraction may have preceded speech seems to be pure speculation, conducted independently of the archaeological evidence. Moreover, Marshack himself weakens this line of speculation by further suggesting (p. 117) that early human utterances always had some direct connection to their immediate surroundings, as in "a cry or specialized word of warning at the presence of carnivore," thus limiting the capacity of these primeval "words" to convey abstractions or even generalizations (e.g., a cry might communicate the immediate presence of a lion, and/or fear on the part of the crier, but would have difficulty, unlike the abstract word "lion," in communicating about the general species of lions, or even any specific lion, when no lion is physically present). The absence of such an abstract language faculty means that an "internal" abstract mentality, if it existed, would have had no convenient mode of expression even after the initial development of speech — which raises the question as to what biological purpose abstract thinking could have served, with no ready means of expression or communication. (As for the possibility of abstract thinking being communicated by a pre-linguistic system such as gestures, Ralph J. Beals and Harry Hoijer point out in *An Introduction to Anthropology* [New York: Macmillan, 1965}, pp. 616-617, that very few symbolic gesture systems are known to exist in human societies, and those that do are always derivatives of earlier oral systems. Abstract gesture systems taught to the deaf are of course also the product of prior oral and written modes of communication.) In sum, then, the only demonstrable and consistent point in Marshack's conjectures about the origins of language is that abstract thinking and speech are as old, or older, than the physical artifacts that Marshack has examined. (This, of course, in no way prejudices Marshack's argument that communication is in part both non-verbal and non-abstract, or the possibility that non-verbal/non-abstract modes of thinking and communicating preceded speech. See notes 16 above and 19 below in this chapter.)

18. Isaac, p. 104.

19. Again, note that what is being discussed here is not "intelligence" or even "thinking," in general, but a particular type of thinking— abstract thinking— that occurs independently of physical surroundings, and has no necessary connection to its referents in the real world (such might also be termed "non-mimetic" thinking, or thinking independent of imagery). As to the existence of mimetic or non-abstract human intelligence before speech, Jaynes, p. 130, suggests that skills needed to make primitive tools could have been transmitted without language, "solely by imitation. . . . It is the same problem as the transmission of bicycle riding; does language assist at all?" Probably to no great degree, for bicycle riding is taught primarily by demonstration and imitation— i.e., replication— not abstract description. See also notes 11 and 13 above in this chapter.

Of course, the lack of evidence by no means negates the possibility that human abstraction did, in fact, prefigure speech and even dictate its development. But such a possibility is best discarded an an explanation for the persistence of speech and abstract media, for it assumes the pre-technology of abstract thinking on criteria much less substantial than that applied to all other patterns of communication thus far identified as pre-technological in the present study. The invention of speech would thus be most reasonably viewed here not as a replication of a pre-existing human mode of abstraction, but as an exceedingly

primitive technological attempt to communicate across time and space at the extreme distortion of the pre-technological environment. In this view, abstraction serves not as an antecedent but as a consequence (or perhaps co-requisite) of technological speech — a consequence, moreover, entirely consistent with the present study's observation that the more primitive the technology, the more profound its departure from the pre-technological reality. But if such is a satisfactory explanation of speech's development, and its relation to abstraction, the explanation appears to once again pose the problem of speech's persistence: why, if speech is a primitive, highly distortive technology, and abstraction its concomitant, have both managed to abundantly survive the advent of immensely more replicative, or less distortive, technologies?

Fortunately, posed in such terms, the problem seems at last soluble within the guidelines of media evolution as discerned in the present and previous chapters. The solution depends, first of all, on the definition of speech and its connection to abstraction developed in the last few pages, which may be summarized in two parts: (1) speech is best considered a technology rather than a non-technology, and a primitive technology that distorts rather than replicates; (2) although speech entails abstraction, there is no demonstrable evidence that abstraction predates speech. Rather, abstraction seems to be the technique that speech uses to extend across time and space, just as chemical emulsion, etc., the technique that photography uses in its extension; but whereas the chemical emulsion preserves part of the pre- technological reality even as it transcends it, abstraction transcends only by completely departing from all elements of pre-technological reality (this includes visual as well as acoustic reality, for with the small exception of onomatopoeia, the words of speech have no literal connection even to the sounds they represent, e.g., "loud," "soft," "laugh," "cry," etc.). This view of speech and abstraction suggests that a characteristic of media evolution first noted in the discussion of media determinism at the end of Chapter 2 above may have relevance to the question of speech's persistence: namely, that the greater a medium's departure from pre-technological reality, the greater the medium's propensity to alter that pre-technological reality, and the humans functioning within that system. Such may be considered a media equivalent of the law of thermal equilibrium: let a glass of water at 70 degrees C , for example, represent the human, pre-technological environment; if that environment is placed in contact with temperatures of 68 or 72 degrees C, i.e., technologies that closely approximate the pre-technological environment of 70 degrees), it will remain relatively unaltered; but if the glass of water is subjected to temperatures of 150 or 0 degrees (i.e., technologies that are at great variance with the pre-technological environment of 70 degrees), its temperature will change profoundly. Since the abstractions of speech are an extreme divergence from the realities of the pre-technological communication environment, the impact of speech upon that environment would be expected to be enormous. And it is in the enormity of this impact that the key to the survival of speech and other abstract media may be found.

What would be the nature of such an impact upon the human, pre-technological system of communication? Coleridge has written that appreciation of poetry is dependent upon "that willing suspension of disbelief for the moment" (Ch. 1, n. 29 above)? but appreciation and usage of abstract speech requires a suspension of disbelief that is soon far from voluntary, and a lot longer than momentary. What both speaker and listener must suspend is the palpably obvious perception that the spoken word is in no way part of the reality it seeks to describe: for speech works only when its words are invested with the full force and effect of the things they describe, only when the pretense that words are the realities they designate is played to the hilt. "My house is on fire" is ineffective communication, unless the listener responds to the neighbor's call as if he smelt the smoke and saw the flames. Thus, assuming that abstraction did not exist beforehand (and, for reasons explained above, such is the assumption here), the perpetuation of speech surely made abstraction a human necessity. In this sense, the suggestion that language was a prerequisite for art and artificial tools may be justified after all, though not quite for the reasons usually given: the debt that art owes language may lie not so much in the use of language as a practical communication system (.though this certainly helped), as in language's serving as a primary raison d'etre for a process of abstraction that, once sufficiently developed and sustained by language, could then be directed toward artistic and other abstract endeavors. Language, in other words, may have literally been the magic word that set loose the genie of abstraction.

But speech itself of course also conferred tremendous and unprecedented practical advantages in communication which, when considered, make it not surprising that this primitive, distortive technology caught on so universally. On the other hand, it is possible that speech was invented numerous times, but died with its inventors owing to the inability of the populace to handle abstractions. At any event, somewhere along the line a critical threshold of sorts must have been crossed (this may have occurred only once or in multiple places and times), and those humans unable to speak — or, to be more precise, who lacked the capacity to generate and understand language — increasingly began to find themselves at a distinct biological disadvantage. For "strong and silent types" to the contrary, speechless people in a world of speakers would have had a hard time seeing to it that their genes made it to the next generation. Even an absurdly conservative estimate of 10,000 years of this relentless selective pressure would allow more than enough time for the capacity to speak, and its attendant process of abstraction, to become genetically embedded in human mentality — to become, in all respects, a fully functioning part of the pre- technological environment. [20]

20. The 10,000 year estimate for the origin of speech is conservative because recent evidence suggests that writing itself is at least that old (see Rensberger, "Roots of Writing Traced Back More Than 10,000 Years," *New York Times*, 9 July 1977, pp. 19, 20). As for the likelihood of such a profound evolutionary change in such a "short" period of time, Wilson, p. 569, points out that "the theory of population genetics and experiments on other organisms show that substantial changes can occur in the span of less than 100 generations, which for man reaches back only to the time of the Roman Empire. . . . Although we do not know how much mental evolution has actually occurred, it would be false to assume that modern civilizations have been built entirely on capital accumulated during the long haul of the Pleistocene." Moreover, the view of language as a special determinant of human evolution has been recently forwarded by Popper and Jaynes in strikingly parallel conjectures. Popper, citing the theory of "organic evolution" developed independently by J. M. Baldwin and C. Lloyd Morgan, explains that "by adopting a new form of behavior the individual organism may change its environment . . . and may thereby expose itself and its descendants to a new set of selection pressures, characteristic of the new environment. . . . We could say that in choosing to speak, and to take interest in speech, man has chosen to evolve his brain and his mind; that language, once created, exerted the selection pressure under which emerged the human brain and the consciousness of self." (*The Self and Its Brain*, pp. 12-13.) And Jaynes, after first arguing that "consciousness is chiefly a cultural introduction, learned on the basis of language," further suggests that "the change to consciousness may have been assisted by a certain amount of natural selection. . . . It is thus possible that individuals most obdurately bicameral [i.e., pre- conscious] . . . would perish, leaving the genes of the less impetuous, the less bicameral, to endow the ensuing generations. . . . Consciousness must be learned by each new generation, and those biologically most able to

learn it would be most likely to survive," (pp. 220-221). (Note, however, that Jaynes insists upon language as a prerequisite for "consciousness," Popper sees language as a prerequisite for "self-consciousness," and the present study argues only that language is necessary for abstraction. Although the distinctions between the three views may be not all that great — in that Jaynes, pp. 21-66, takes pains to define consciousness very narrowly, and as operating through "the analogue 'I'," and Popper, p. 444, suggests "human consciousness of self" is "based on abstract theories" — the linkage of language with any type of consciousness (as opposed to abstraction) inevitably reduces the possible range of consciousness for animals and, perhaps even more unjustifiably, for pre-linguistic children. Indeed, although the abstractions of language no doubt play a large role in the refinement and perhaps even construction of consciousness of self, such consciousness may very well also be present in the very concrete, and pre-linguistic, perception of hunger, or in the infant's perception of "itself" as distinct from the surrounding environment. Thus, the present study argues only that language is necessary for abstraction, and those aspects of consciousness and self-consciousness (admittedly large, but not necessarily total) that depend upon abstraction.

Thus, the survival of speech has indeed ultimately been due to its reflection of an aspect of the pre-technological environment— but an aspect of its own making. This was possible because speech was for so long the only decent communication game in town that eventually its distortions became expected, normal, "natural." Of course, the change in the pre-technological environment was not so profound as to obliterate the desire for non-abstract media (the thrust of this entire study argues the opposite), but the alteration was deep enough to assure the continuance of speech and its abstract derivatives once the non-abstract, replicative media finally began arriving in force in the 19th and 20th centuries. Indeed, so deep was the ecological niche that speech had carved out for itself that, long before the arrival of replicative media, a whole genre of purely abstract content had arisen —abstractions which not only had no literal

connection to the real world, but no connection or referrent whatsoever in the real world, and which were transmittable only by abstract media. Thus, while the word "tree," for example, represents (albeit arbitrarily) something which can be seen, touched, as well as abstracted, where in the real world could one hope to find and touch what is represented by the word "concept," and how would one go about photographing it? It is difficult to even imagine the existence of such abstractions before the development of language — because, being uncommunicable, what purpose would they have served? — but their existence since speech, in any case, has given this and other abstract media a monopoly on their communication, a niche unencroachable by media that replicate the real world.

Of course, competition for the niche of abstraction has often ensued among the abstract media themselves. As Innis and McLuhan have so thoroughly documented, phonetic writing and its extension through print have at various stretches in history tended to eclipse the importance of speech (though speakers and listeners even in the most print-dominated age surely outnumbered the writers and readers). But the primacy of speech, and its proximity to the pre-technological environment, is reflected by the fact that it has in one way or another become the content of every successful replicative medium — including even the computer — with the exception of still photography. In the long run, the double abstractions of phonetic writing proved to be over-extensions not entirely supportable even by the new niche of abstraction, and writing and print thus proved the more vulnerable to incursions by photographic and electronic technologies.

Although television has been frequently cast as the "enemy" of print — by McLuhan, for example[21] — the present analysis suggests that print has the most to fear from media that permanently record speech. Since print (writing) is still a highly effective communicator of abstractions, print has an ecological advantage over all replicative media — save those that replicate speech. Moreover, print has an advantage over unrecorded speech in that print's abstractions are more easily extended across time, or permanentized. Thus, print can be outflanked only by a medium that provides a permanent transcription of the spoken word — that is, a medium which competes with print both in abstraction and permanency. Television provides content that is mostly non-abstract (pictoral) and, till now, non-permanent; telephone speech, like non-recorded speech, is evanescent. It is thus Edison's talking machine, and its descendant the cassette tape recorder, that are most likely to be print's ultimate undoing. [22]

21. McLuhan contends that television, unlike photographic and other electronic media, poses an actual physiological threat to reading, in that "the instant flow of TV imagery tends to immobilize the motor muscles of the eyes, creating sleepiness and also impeding the eye movements necessary for reading print. . . . The very antivisual effects of TV are the crux where literacy is concerned. Movies did not inhibit the motor responses of eye muscles." (McLuhan, review of *Television: Technology and Cultural Form*, by Raymond Williams, *Technology and Culture* 19 [April 1978]: 261.)

22. It is interesting to note m this connection the report of Hiebert, Ungurait, and Bohn, p. 316, that the public spends "almost as much [money] on records as it does on textbooks" — an impressive comparison, in view of the fact that most textbook sales are "forced" (i.e., required of students), whereas most record sales are voluntary.

Several recent developments, however, suggest that the future of print may not be all that grim. A survey conducted in Indiana contradicts the widely held view that reading ability has deteriorated in the United States during the past 30 years.[23] Television itself, seeking new types of programming to meet the possibilities of cables and computers discussed in the last chapter, has been experimenting with news "print-outs" as an alternative to the standard TV news format — this beginning to fulfill the long-standing prediction of "televised" newspapers.[24] In the same vein, the huge growth of Xerox and similar print photocopiers, and the introduction and apparent success thus far of fascimile mail (writing transmitted by electronic signals over telephone wires and recorded by impact printers, Xerox, or any of a variety of possible technologies), suggests growing areas of cooperation between print and replicative media.[25] Indeed, it is possible that having at first stolen print's thunder, electronic and photographic media are now beginning to retrieve print as their content in much the same way as they previously retrieved speech.[26] This suggests that after thousands of years of exposure, the peculiar abstractions of phonetic writing may have gotten into our blood after all (if perhaps not yet our genes), and may very well endure, though in a capacity far diminished from that of their heyday in the 18th and 19th centuries. As for the future of paper books and newspapers, however, this remains far less certain.

23. See Gene I. Maeroff, "Reading Achievement of Children in Indiana Found as Good as in '44," *New York Times*, 15 April 1978, p. 10. Although the survey was conducted in Indiana, Maeroff reports that researchers Leo Fay and Roger Farr see no reason that their findings need not have national applicability.

24. This development is most advanced in Britain, where "Ceefax" and "Oracle" systems have been servicing subscribers with televised printed news for several years. In addition, the newer "Prestel" system, introduced in Britain on June 1, 1978 and expected to be available there nationwide by 1979, hooks subscribers' telephones to a central computer system, and allows them to dial for a variety of printed information for display on their home TV screens. See Nigel Hawkes, "British May Use

Telephones, TV's, to Tap Data Bank," *Science*, 9 July 1978, pp. 33-34, for brief descriptions of the Ceefax/Oracle and Prestel systems; see Maddox, p. 36, for a brief review of the history of televising printed material. It is also interesting to note in this regard that in New York City, WNBC-TV uses a printed "Data Bank" of information as a bridge between its live news reporting and commercial breaks.

25. See Robert J. Potter, "Electronic Mail," in *Electronics: The Continuing Revolution*, eds. Abelson and Hammond, pp. 90-94, for a discussion of various types of fascimile mail systems, and an analysis of how they would perform traditional mail functions. Another significant kind of cooperation between print and photo-electronics is evidenced in the *New York Times*' recent shift from a mechanical Linotype printing system (in which trays of lead type were used) to a photocomposition system involving computer terminals, laser scans, and photosensitive paper. This new system not only produces 1,000 newspaper lines a minute (a task that would require the simultaneous labor of 200 Linotype operators), but allows for the fascimile broadcasting of all material initially typed (or electronically encoded) into the computers — thus bringing the New York Times a big step closer to the televising of newspapers discussed in note 24 in this chapter above. See Malcolm W. Browne and Carey Winfrey, "*The Times* Enters a New Era of Electronic Printing," *New York Times*, 3 July 1978, pp. 21, 38, for a full discussion of the change at the *Times*.

26. Although the use of electricity to communicate initially competed with print, and did so for a long time thereafter, the use of electricity for lighting purposes was from the very start a powerful enhancer of print, as David de Haan points out in *Antique Household Gadgets and Appliances* (Woodbury, N.Y.: Barron's, 1977), p. 121, when he writes that "electric lighting did more to improve the habit of reading books than anything before it." In the sense that books and newspapers read under electric lights were, in effect, being in part "transmitted" by those lights, one could argue that print has been the "content" of electronic systems for a good deal longer than televised newspapers and fascimile mail.

The process of abstraction, first surfacing as a tool of

speech, has thus cut a swath both wide and long in the pre-technological environment. Through a kind of "functional autonomy," to borrow a phrase from psychologist Gordon Allport,[27] the symptoms of abstraction have in many ways far outshone their original incubator of speech, which there seems little reason to believe was any more than a crude, if ingenious, attempt to transcend time and space in the utter absence of any other technologies to better do the job. It is intriguing, therefore, to wonder what would have become of the human race had our speechless ancestors been given a present of a Polaroid camera. Perhaps much the same thing as would have happened to our legs had we been put in an automobile before we learned to walk upright.[28] Yet it is interesting to note that it took hands freed by upright legs to build the automobile, just as it took abstract thoughts set loose by speech (and writing) to invent the camera. Thus, the evolution of media whose processes are successively less abstract is, in the most profound of ways, the consequence of media whose processes were successively more abstract.

27. Allport introduced the term in *Personality* (New York: Holt, 1937) to describe behaviors that may become separated from their original motives, yet continue on their own momentum, or by developing motivations of their own. A miser, for example, may have begun hoarding money to overcome poverty, yet continues hoarding money "for its own sake," long after the poverty motive has gone.

28. The "see-saw" between generalization-through- abstraction and generalization-through-replication/extension has already been suggested in Ch. 4, n. 17 above, and is described more fully by Arnheim, who writes, "We must not forget that in the past the inability to transport immediate experience and to convey it to others made the use of language necessary and thus compelled the human mind to develop concepts. For in order to describe things one must draw the general from the specific; one must select, compare, think. When communication can be achieved by pointing with the finger, however, the mouth grows

silent, the writing hand stops, and the mind shrinks." ("A Forecast of Television," reprinted in *Film As Art*, p. 195). The severity of the last sentence perhaps reflects the fact that, writing in 1935, Arnheim had no way of knowing that abstract communication would survive the first 40 years of televised "finger-pointing" quite handsomely, as described in the preceding discussion.

***

Perhaps the most useful lesson to be learned from the above attempt to explain the persistence of abstract media within the framework of anthropotropic theory is that the pre-technological environment, used thus far as a standard upon which to gauge the evolution and survival of media, is in fact only as standard or immutable as genetics. The pre-technological standard of color is thus unaltered by 100 years of black-and-white photography, but the pre-technological world of concrete realities is mightily altered by 10,000 years or more of abstract language, which indeed erects a new standard of pre-technological reality. The first observation is this theory's reason for rejection of most of media determinism; the second is this theory's recognition of the fact that in the long, long run, even the most constant aspects of human nature — any nature — dance to the tune of changes in the environment. Since the predictions and scenarios of future media to be attempted in the following chapter will in effect be projections of how the retrieval of the pre-technological environment will be completed, these predictions must be tempered with the understanding that the pre-technological environment itself may someday not be what it is today. Moreover, as tentative as some of the explanations of past media development have thus far been, the attempts to describe the future will be even more so, since these obviously deal in events that haven't even occurred as yet. Nonetheless, as suggested at the outset, a theory is obliged to make predictions precisely

because of the possibility of being proven wrong. Having thus acknowledged its limitations and responsibilities, the present theory will now go on to hazard several guesses about the future evolution of media — guesses which, being informed with all that has been observed thus far about media evolution, will perhaps stand up to the unblinking eye of time.

CHAPTER 7
FUTURE REUNIONS

MOBILITY AND ACCESS: WIRES, AIRWAVES, AND LIGHT BEAMS

One of the most noticeable shifts in 19th century to 20th century technology in all areas has been the supplantation of fixed, rigid systems by flexible, often physically unconnected components. In transportation, this change has been especially obvious, with automobiles replacing railroads, buses replacing trolleys, and of course both up-staged by the growing ascendancy of air travel. (Transportation, as will be seen below, enjoys a special relationship to anthropotropic media, and it is therefore not surprising to note that in the pre-technological world, transportation occurs on flexible rather than fixed routes — suggesting that the latest transportation advances are recapturing the pre-technological reality of transportation.) In the realm of media, as described in previous chapters, this trend towards flexibility was evidenced first with the introduction of broadcast technologies as improvements over wire technologies/ and more recently with the complete portability of radio and TV receivers, as well as recorders, cameras, and some transmitters, made possible by miniaturization of batteries, transistors, and computers.[1]

1. "Pocket-size" televisions have been available to the U.S. public since early 1978, see J. S. & A's *Products That Think* 1977-1978 sales catalog p. 38 for description of a model that sells for $199. As for the increasing flexibility of recording and transmission systems, Monaco, p. 79, concludes that with recent innovations "the camera approaches the ideal condition of a free-floating, perfectly controllable artificial eye," and cites the filming of the inside of the human body in the recent documentary *The Incredible Machine* (broadcast in the Fall of 1976 in the U.S.) as evidence. For a technical discussion of the new microcomputers (made of components so small that 400,000 fit to the square inch) which make many of these developments possible, see Raymond E. Dessy, "Microprocessors? — An End User's View," in *Electronics*, eds. Abelson and Hammond, pp. 138-145.

Since communication within the pre-technological world is completely flexible and untethered — with any individual able to dispense and receive information from any area within physical proximity at will — the wireless, portable evolution of media should continue to the point of providing any individual with access to all the information of the planet, from any place on the planet, indoors and outdoors, and of course even beyond the planet itself as communication extends into the solar system and cosmos beyond. Radios in automobiles, and movies and TV in airplanes, are but the modest beginnings of such a "systemless" system which will eventually give the individual the same unrestrained access to information on a global basis that the individual has always enjoyed to information in the immediate physical environment.

A problem arises, however, in that the technology that has thus far provided the most mobility in communication, has not been able to provide the most variety or even clarity of information available. Thus, the tenuous electromagnetic wavelengths used in radio and TV broadcasting are, as already indicated in Chapter 5 above, subject both to interference and to absolute physical limitations on the amount of information — i.e., number of channels — capable of broadcast; while the older, fixed wired systems offer minimal interference and are limited only by the size and number of wires constructed. In effect, then, at least in terms of capability for expanding the variety of information delivered, broadcasting would seem a more "fixed" system than wires — since it is certainly more difficult to add new usable frequencies to the electromagnetic spectrum than it is to build new wires. Thus, a situation of universal access to universal information would require a new technology that goes beyond broadcasting and wires, and combines the mobility of broadcasting with the expansion capacity of wires.

Attempts at current cable technology as an alternative to broadcast television are no doubt motivated by a desire for such increased capacity; but these cables are even more cumbersome than conventional wires (they are much heavier) and as such impose a level of immobility that would probably prove an unacceptable trade-off to a world that has already tasted the freedom of flexible broadcasting. Moreover, co-existence of cable along with broadcast technologies as a way of providing increased capacity where needed (in the home, for example, but not the car) runs contrary to the principle of convergence — that is, one unified system rather than several cooperating systems — that has already been noticed as an emerging pattern of media evolution. Cable television is thus not a likely candidate for long-range survival.

The fiber optic technology, however, briefly mentioned in Chapter 5 above, seems much more promising. As already indicated, fiber optics deliver information on pin-point light beams or lasers guided through almost as minuscule glass fibers. Even such an initial system as that currently under development at Bell Laboratories[2] would reduce the bulkiness of cables to a fraction while increasing astronomically the delivery capacity over both broadcasting and cable. But the flexibility of a light-beam system of communications far exceeds that of even tiny glass fibers: for as Arthur C. Clarke has recently pointed out, lasers need not be confined to glass conduits; they are capable of transmission, in theory at least, around corners, over and under obstacles, and to any part of the universe through the use of mirrors.[3] Thus, laser communication — currently being developed in only its most primitive form as fiber optics — seems the ideal technology to at last provide the combination of mobility and scope that has beckoned media since the invention of the radio. It is a safe anthropotropic bet, then, that the laser will replace both the broadcast that is flexible but limited, and the cable that is expansive but rigid.

2. See W.S. Boyle, "Light-Wave Communications," *Scientific American* 237 (August 1977): 40-48, for a description of Bell's first commercial testing of fiber optic telephone service in Chicago.

3. Arthur C. Clarke, "Communications in the Second Century of the Telephone," in *The Telephone's First Century — and Beyond*, with a Preface and Afterword by John D. deButts, and an Introduction by Thomas E. Bolger (New York: Crowell, 1977), pp. 98-99. Clarke further suggests that "leakage" of light waves from glass pipelines be encouraged, as a way of providing access to information not only at the end of the pipeline, but to everyone along the way.

OBSERVATIONAL TERMINUS: HOLOGRAPHY AND THE RETRIEVAL OF THE THIRD DIMENSION

But what would be the content of such a distribution system that can reach anyone, anywhere, with anything? Media, as described in previous chapters, have thus far managed to replicate the forms of the audio-visual pre-technological world in considerable detail and fidelity — the one significant omission being the visual third dimension of depth. That the desire to view things in their entirety — from the sides, from above and below, from behind as well as just from the front — has been as strong as other desires for media, now satisfied, to view replicated life in motion, color, and so forth, is demonstrated by a brief passage from a science fiction story written by Isaac Asimov in the 1940s, several years before even the most rudimentary three-dimensional technology would be possible:

> In five million homes on Terminus, excited observers crowded around their receiving sets

> more closely.... The center of the chamber was cleared, and the lights burnt low.... and with a preliminary click, a scene sprang to view; in color, in three-dimensions, in every attribute of life but life itself.[4]

The technology that now seems most likely to provide the third dimension of visual life is, most interestingly, a type of laser photography called holography — thus furnishing another instance of media convergence; or one technology, in this case lasers, responsible for both the distribution and replication of three-dimensional images.

[4]. *Foundation* (New York: Avon, 1951), p. 186. Note that the chapter from which the above quotation was taken ("The Merchant Princes") was first published as a short story in the early 1940s, see Joseph F. Patrouch, Jr., *The Science Fiction of Isaac Asimov* (New York: Doubleday, 1974), pp. 62-63.

The inability of current broadcasting technology to transmit holograms has already been discussed, as has been the solution to the problem through fiber optics (see end of Chapter 5 above). It should be noted, however, that even as a replicative technology, three-dimensional holography is far from perfected. The reproduction of a moving object in three dimensions, or even a still object in a complete 360-degree panorama, is currently accomplished by superimposing a series of individual holograms, which record the moving object in successive positions or the still object from successive angles, upon a single photographic plate; the object "reappears" in three dimensions when a laser is beamed through the plate: but the condensation of multiple impressions upon a single plate often produces an image that is far less clear than that of a standard two-dimensional photograph or motion picture.[5] In a forthcoming dissertation on changing conceptions of visual space, Ed Wachtel of New York University suggests that the problem with current holography may lie in its attempt to create a total three dimensional image by in effect combining images that are only partially three-dimensional, rather than attempting to somehow capture an image that is from its inception totally three-dimensional, in much the same way that motion pictures are both projected as well as initially filmed in motion.[6] Examination of the history of motion pictures suggests that Wachtel may be correct, for the early attempts to reproduce motion by taking a series of isolated still photographs and projecting them in motion, produced unlife-like, jerking pictures — only by capturing as well as projecting the images in motion did the illusion begin to approach reality.

5. Holograms of still objects in substantially less than a complete 360-panorama can be produced by a single impression upon the photographic plate, and are thus a good deal clearer; yet in failing to replicate both motion and multiple perspectives, such holograms capture only a limited segment of the three-dimensional world. See Stroke, p. 222, for a brief further discussion of problems in motion holography.

6. *The Transformation of Visual Space: A Theory of the Relationship Among Technology, Space Conception, and Culture Change* (Ph.D. dissertation, New York University, forthcoming).

But whether the replication of the third dimension comes from holography or a completely different technique, there can be no doubt that, once achieved, the three dimensional will replace the two dimensional image as surely and completely as color replaced black-and-white, and talkies replaced silent movies — for unlike the stillness of the real world which was accurately captured by the early medium of still photography, there is nothing in the pre-technological world that is two-dimensional, and can't be better replicated by a three-dimensional medium.

But what of sensations other than sight and sound, such

as smell and touch? Media such as "smellovision" and "Sensurround" that attempt to replicate these sensations usually seem awkward and almost comical, and with good reason. For in the pre-technological mode of observation (as opposed to interaction, see immediately below), senses other than sight and sound play very small roles: in the case of observing the moon or anything far off, smell and touch serve no function at all (a miner exception would be "feeling" the warmth of the sun); and even in other situations of observation rather than interaction, touch is rarely used. Thus, in the realm of observational media at least, technologies that go beyond sight and sound will probably remain only as playful gimmicks.

But communication in the pre-technological environment is often not observational but interactional. And in conversations between two or more people, touch and other non-audio-visual perceptions may play significant roles. Thus, the media that attempt to approximate pre-technological interactive communication may go well beyond the third dimension of holography.

THE AGING HEIR APPARENT: VIDEOPHONE

The most advanced interactive technology thus far developed has been the videophone or picture telephone, which has been on the verge of entering the mainstream of communication for nearly twenty years, or perhaps longer than any other new technology.[7] Ironically, once the videophone does become available for mass use, it will probably replace the telephone in a much more total way than television replaced radio (for, indeed, as discussed in the previous chapter, television has not really replaced radio at all). Radio survived the advent of television, as discussed in the previous chapter, because radio approximates the pre-technological communication pattern of observational hearing without seeing, or eavesdropping. But whereas people in the pre-technological world may hear or listen without sight — as in being wakened from sleep by sound — they rarely converse or interact without sight. Indeed, the studies of body language suggest that non-verbal perceptions play an essential role in face-to-face and other forms of interactive communication in the real world.[8] Thus, faced with the choice of blind telephone or sighted videophone, people should — if this anthropotropic analysis is correct — choose the videophone (unless, of course, they have a special need for privacy, in which case they would simply turn the video portion off, or perhaps put up a smiling photograph in place of the live image; see Chapter 9 below for discussion of options in anthropotropic technologies).

7. Maddox, pp. 207-208, captures the mixed career of the picture telephone when she writes: "If any single piece of the new communication technology has been advertised in advance, it is the picture telephone, yet there is little sign of any demand for it at all. Telephone manufacturers trot out a picture telephone whenever a world-of-tomorrow exhibit is called for. . . . Perhaps the demand will materialize..." Conversing with pictures has been made possible thus far both through Bell Telephone's "Picturephone," described by Gregor, pp. 81-91, and through interactive cable television, currently being tested in Reading, Pa., and described by Gregory Jaynes in "A New TV Idea: Shows for Old People," *New York Times*, 7 July 1978, p. A7. (See also Ch. 1, n. 8 above.)

8. The Communications Study Group in England, for example, cited by K o r z e m y pp. 16-17, reports that "recent film analyses of conversations have provided striking evidence of the almost ballet-like inter-weaving of looks, gestures and body movements, as successive speakers take the floor, pause, are interrupted and resume talking. It follows that if the visual channel is removed, and the discussants can no longer see each other, this synchronization should be impaired." And Korzenny, p. 17, himself concludes that "the extent to which we feel close to the others should be a function of the resemblance of the mediated interaction to the traditional [i.e., non-mediated or real-world] interpersonal encounter." See also Paul Levinson, "'Hot' and 'Cool' Redefined for Interactive Media," *Media Ecology Review* 4 (February 1976) : 9-11, for a discussion of the importance of maximum information delivery in interactive media systems.

But such being the case, why has videophone taken so long to be publicly accepted? Or, asking the question from the perspective of the radio and television example; why was observational radio, which at least successfully replicates some aspect of the pre-technological environment/ supplemented with observational television; while interactive telephone, which in effect fails to completely replicate any aspect of the pre-technological world, was left so long alone? The answer to both questions is the same. Certainly, it has not been for any lack of anthropotropic desire on either inventors' or consumers' parts that the videophone has failed to publicly materialize: indeed, the first public demonstration of television in the 1920s was as an interactive medium,[9] and telephones with picture screens have long been a part of the public's collective imagination, as evidenced in science fiction themes, television shows, etc.[10] But the problems of economics/ politics, and inadequate transmission technology were overwhelming. Having a television screen in every home allows for observation of sounds and images; but interactive video requires not only screens or receivers, but cameras and transmitters which are far more expensive. Moreover, electromagnetic broadcast technology is not well suited for interactive communication — thus it fell to the wires of the telephone to do the interactive video job. But this in turn created two new problems because (a) existing telephones are designed for the lower informational transmission of sound rather than picture sources, and (b) U.S. government regulations forbid the telephone company from any involvement with existing television systems, which presumably includes the perfectly adequate screens that are now in people's homes. The telephone company was thus forced to introduce "Picturephone" on special public screens, which offered such awkward and limited access that they ware bound to fail.

9. Gregor, pp. 81-83.

10. "Well, I want to arrange to have two kinds of pictures come over the wire. I want it so that a person can go into a booth, call up a friend, and

then switch on the picture plate, so he can see his friend as well as talk to him. I want this plate to be like a mirror, so that any number of images can be made to appear on it. In that way it can be used over and over again. . . . No matter how far two persons may be apart they can both see and talk to one another. . . . Then another thing I want to do is have it arranged so that I can make a photograph of a person over a wire." — This spoken by Tom Swift in Victor Appelton's *Tom Swift and his Photo Telephone* (New York: Grosset & Dunlap, 1914), pp. 131-132. More recent examples of videophones in the public's imagination are comic strip character Dick Tracy's special wrist-watch (which has a two-way TV screen on its face), and the "main screen" on the "bridge" of the Star Ship Enterprise on TV's *Star Trek* series.

Cable television — which provides better transmissional technology than conventional telephone lines, and is relatively free from government supervision — has alleviated some of the obstacles to interactive video; and indeed the first interactive video systems in general use, utilizing the cable technology, are now operating in Reading, Pennsylvania and several other cities in the U.S.[11]. But for reasons of inflexibility already discussed earlier in this chapter , cables will probably perform only temporarily in the ultimate ascent of videophone. For much the same reasons, a far more satisfactory conduit of videophone would be the fiber optic and then the free laser technologies discussed above, which would provide both flexibility (i.e., the ability to make a videophone call from your pocket) and sufficient intensity of energy transmission to clearly communicate pictures as well as voices. And lest these lasers begin to sound like the magical panacea for all anthropotropic problems, let it be recalled that Alexander Graham Bell first transmitted voices on a beam of light with his "photophone" in 1880, a device which was, in the inventor's own words, "the greatest invention I have ever made; greater than the telephone."[12]

11. Note, however, that, similar to the initial offering of Picturephone service, the "interaction" in these interactive cable systems thus far is limited to public studios, rather than the home. In the Reading system described by Gregory Jaynes, p. A7, for example, the public must go to a local TV station to converse with officials, etc.via TV (these conversations, however, are transmitted via cable to subscribers who can watch them on their home TV screens — but the home viewers cannot join in the conversation).

12. "What's next, Professor Bell?", advertisement appearing in New York Review of Books, 28 October 1976, p. 8.

And, of course, the same lasers that transmit videophone would be able almost as easily to transmit voices with three dimensional images, which might be termed "holographone." And this lasers should someday certainly do.[13] Yet even the holographone would pale in comparison to what the conclusion of interactive media evolution might someday bring.

13. Many of the appraisals and projections of the last few pages have already been suggested by Clarke, pp. 93, 96, who writes, "I am aware that previous attempts to supply vision — such as the Bell Picturephone — have hardly been a roaring success. But I feel sure that this is due to cost, the small size of the picture, and the limited service available. . . . Such technical limitations have a habit of being rather rapidly overcome, and the *large screens high definition* [italics in original] Picturephone-Plus is inevitable. . . . It is usually assumed that the console would have a flat TV-type screen, which would appear to be all that is necessary for most communication purposes. But the ultimate in face-to-face electronic confrontation would be when you could not tell, without touching, whether or not the other person was physically present; he or she would appear as a perfect 3-D projection. This no longer appears fantastic, now that we have seen holographic displays that are quite indistinguishable from reality. So I am sure that this will be achieved someday . . . " (On the issue of mobility, Clarke, p. 97, also suggests that the telephone will

"lose its metal umbilical cord" to become "the individual, wristwatch telephone through which you can contact anyone, anywhere"; the transmissional technology for all of the above would be lasers, Clarke, p. 98, and see note 3 above in the present chapter.)

INTERACTIVE TERMINUS: THE REUNION OF TALKING AND WALKING

To see the future of these media that evolve towards human environments, we must, as always, return to examine the human environments themselves. In the pre-technological world, there is no difference between communication and transportation. ..They are the same act — for to give someone a mile away a message, one must, if unaided by any technology, walk into that person's physical presence. The invention of primitive technologies changed that: long before phonetic writing's fragmentation of the visual from the other senses led to the rise of logic, science, and civilization, it cleanly severed the act of communication from transportation as surely as ying from yang, or Eve from Adam. From that day on, people could send and receive information without sending and receiving themselves. But like the two halves of their Chinese and Biblical metaphors, communication and transportation have maintained a continuing affinity — the obsolescence of wires and railroad tracks alluded to at the beginning of the present chapter being just one example of the communication/transportation evolutionary parallel.

Perhaps less obvious has been the gradual, actual convergence of these two pieces of the same whole. In communication, the convergence began first with the sending of information faster and faster until at instantaneous speed, and the then gradual sending of more and more of the human body — first the sound, then the image, and perhaps soon the three-dimensional image — at that speed; in transportation, the process began with the sending of the entire physical human body, first at slow but now faster and faster speeds. Thus, as communication successively humanized its speeding message from telegraph to telephone to videophone, transportation successively sped its already fully human "message" from horseback to auto to jet.

With holography and fiber optics on the verge of instantly transmitting three-dimensional color, motion, and talking images, and a new magnetic propulsion system being seriously proposed which would transport people from New York to Los Angeles in seven minutes (!);[14] it is plain to see that communication and transportation are well on their way to once again becoming one and the same. Whether the final breakthrough comes from communication or transportation is irrelevant; whether the ultimate device operates like a *Star Trek* transporter is also irrelevant; what does matter — and what now seems certain — is that the end point of the evolution of interactive media would allow anyone to be instantaneously and physically present anywhere on the planet that person was welcome (for there would, after all, have to be someone at the receiving end to turn the receiver on) . (Note, by the way, that this could work only on an interpersonal, one-to-one basis, since one physical body can only be in one physical place at a time —

that is, Walter Cronkite couldn't physically be in every home even if we wanted him to be — which we probably wouldn't, since we observe rather than interact with newscasters, or enjoy watching them but wouldn't want them to watch us.)

14. Reported on "Eyewitness News," WABC-TV, 13 February 1978, between 6 and 7:00 PM. According to Henry Kolm, "An Electromagnetic 'Slingshot' for Space Propulsion," *Technology Review* 79 (June 1977): 61-62, "electromagnetic flight has long been proposed for highspeed ground travel on earth. Wheel-less vehicles, suspended and propelled by magnetic forces along a metal guideway, could travel at speeds up to 300 miles per hour, limited only by air friction. If vehicles were run inside an evacuated tunnel, speeds would be essentially unlimited. Budgetary squeezes and unproven economics, however, have kept electromagnetic flight on earth at the Wright Brothers stage." (The estimated cost of the system reported on WABC-TV was 250 billion dollars.) See the rest of Kolm's article for a further description of magnetic propulsion; see also Harold M. Schmeck, Jr., "Scientists Talk On Solar Power Stations Aloft and a Super Subway," *New York Times*, 14 February 1978, p. 12, for a discussion of the magnetic propulsion system proposed by Dr. Robert M. Salter of Rand Corporation (apparently the source of the WABC-TV nev/s item).

When this ultimate — but entirely plausible and supportable —communication/transportation technology is perfected, the human body will once again be united with the spirit, but this time with only the spirit's limitations rather than the body's. At least as far as distance and space are concerned (for time still seems impenetrably closed to travel and interactive communication through technology), only imagination — or lack of it — would place boundaries on our ability to fully and physically communicate.[15] When this occurs, the promise of technology so well understood by thinkers from Freud to Teilhard will at last have been achieved: we will truly be gods on our own planet, and perhaps beyond.

15. For this reason, the view that communication will (or should)

ultimately "replace" transportation — as in business deals, for example, being consummated via videophone rather than personal contact — must be regarded as somewhat shortsighted. Arthur C. Clarke's advice, p. 91, to "don't commute — communicate!", for example, overlooks the likelihood that commuting and communicating will someday be the same act.

CHAPTER 8

SUMMARY

At this juncture, the anthropotropic theory of media evolution temporarily rests its case. In general, this theory has argued that there is a pattern in the invention and survival of communications media, an underlying logic and sequence in the development of media which may be summarized as follows:

In the beginning, humans communicated without technologies or artificial contrivances. In this "pre-technological" world, humans saw in colors and in three dimensions, and heard sounds emanating from multiple sources, usually synchronized with sights. Humans had access to all information present in their immediate surroundings, and had complete mobility within these surroundings — that is , humans were able to receive and dispense information at will from any place within their immediate physical environment. But this communication could occur only within immediate physical surroundings, for communication without technology was restricted by the biological limits of eyesight and earshot, and extended only through the human process of memory. This situation persists to the present day, whenever we communicate without technology.

Technologies were invented in order to overcome these limitations. When a word is written on a piece of clay or papyrus, that word can be read by eyes hundreds of miles away, or hundreds of years later. Events described by words are thus no longer tied to immediate physical surroundings. Humans who use words and other primitive technologies to communicate are thus able to perceive events well beyond their physical boundaries.

But real events described by words, or conveyed by other primitive technologies, are in many ways not at all like the original events themselves. Written descriptions of events have no physical dimension, no color, no sound, no motion or even stillness. This situation persists also to the present day, whenever we use words or other primitive technologies like the telegraph to communicate.

Further technologies were invented to overcome these limitations of prior technologies. Photography transmits events well beyond their physical surroundings, without sacrificing the literal, visual form of the event or object. Subsequent developments in photography restored motion, sound, and color to communication beyond physical boundaries. Electronics provided a speed and interaction of communication on a global basis that parallels the speed and interaction of communication in immediate physical surroundings. The process continues in the present day, with new technologies such as holography restoring the third dimension to events, and fiber optics promising an even more universal, unrestricted distribution of these events.

The present theory, then, sees the evolution of media as a gradual and ongoing change from media that extended events across time and space by distorting the events, to media that extend events across time and space with successively less distortion. In other words, media create environments which increasingly recapture the colors, sounds, or "content," as well as the access, mobility, or "process," with which human communication began. Media attempt to make information available on an extended or global basis with the same access, mobility, and real-life sight and sound that characterizes the availability of information in immediate physical surroundings. These characteristics of "pre-technological" communication serve as the basis for predictions of future media developments, and explain developments in media that have already occurred. The theory suggests that media survive based upon how well they approximate the content (colors, sounds, etc.) and process (access, mobility, etc.) of pre-technological communication.

Radio, for example, survives because in presenting sound without sight it approximates the common pre-technological pattern of hearing without seeing, which can occur merely by closing our eyes; silent movies, on the other hand, have all but disappeared because seeing without hearing rarely occurs in the real world. Media such as silent movies which fail to correspond with some element of pre-technological communication are vulnerable to competition from media that do: the "talking" motion picture thus proved overwhelming competition to the silent movie. In effect, media compete with each other for survival in much the same way as organisms compete for survival in the biological world, but the selecting environment in the case of media is the human preference for media that recapture elements of human communication.

Predictions of future media developments follow from this principle and its corollaries. These predictions cover three general areas of media operation: mobility of media users, access of media users to information, and content of the media transmissions.

In the area of mobility, the theory predicts the advent of media that are completely portable and detachable, since in the pre-technological world communication is never dependent upon rigid, fixed systems such as wires.

This means that current wire and cable technology must eventually give way to more flexible technologies, such as those promised by new developments in fiber optics.

In the area of access, the theory predicts that every media user will have the ability to contact any other media user and all recorded sources of information at will, in much the same way as any individual in the non-mediated real world has complete access to everyone and everything within immediate physical proximity. The implementation of such access on a global basis will require two types of technology: (1) a system of distribution that allows for far greater volume and variety of information than either current broadcasting or cable technology (fiber optics may, again, be able to move such quantities of information); and (2) a system for storage and organization of all information, such a system probably to be provided by some type of computers. In effect, the combined result of completely mobile media users having access to all possible information gives every individual a potentially universal access to universal information — that is, all information available to anyone, anytime, anywhere.

The content of such a system — that is, the nature of the information transmitted — will reflect the varied character of human, pre-technological communication systems. To the extent that abstract thinking has become a part of human mentality, the ultimate media system will make available abstractly presented information in the form of computer print-outs on screens and vocal read-outs on speakers. When literal observation of events is required, the system will provide images that are fully three-dimensional, and virtually indistinguishable from the original. In the case of interactive communication, there will in fact be no difference between the original and the representation, for rather than communicating a speaker's three-dimensional image, the ultimate "telephone" will communicate the speaker himself — that is, the "caller" will be instantly and fully transported into the physical presence of anyone the caller wishes to speak to, and who wishes to receive the caller. This prediction follows from the pre-technological situation, where all conversations entail the physical presence of both parties.

The limitations of these predictions arise both from weaknesses in the theoretical constructs themselves, and from variables outside the range of the theory. The theory predicts that media will increasingly recapture elements cf the pre-technological communication environment, but an inability to precisely define the pre-technological environment itself reduces the certainty of these predictions. Is speech, for example, a "technology" or a "pre-technological" mode of communication? The answer to this question has bearing on how well abstract language can be expected to compete with more literal media such as holography in the future. Moreover, whether clearly defined or not, the pre-technological environment has been seen to change in response to prolonged exposure to primitive, or distortive, technologies. The theory speculates, for example, that the long usage of speech and even phonetic writing has naturally selected humans with a predisposition for abstractions, thereby making abstract thinking a part of the pre-technological environment. To the extent that the pre-technological environment is itself a variable subject to change and not a constant, predictions based upon the pre-technological environment are commensurately uncertain.

Variables outside the evolution of media which tend to limit the predictions of the theory include economics, social custom, political regulations, and similar cultural factors. The advent of the videophone, which the theory predicts as providing a more pre-technological type of face-to-face communication than the telephone, is a good case in point. Video-phones in every home require the presence of video-cameras in every home, which are currently a good deal more expensive than video-receivers or television screens. Moreover, the presence of a camera in your home that is hooked into a public system raises the issue of invasion of privacy, or of people seeing you when you don't want them to. Finally, U. S. anti-monopoly regulations forbid any telephone company involvement with any communication involves television. All these factors have worked against the replacement of the telephone with the video-phone which, as suggested above, seems a logical development on the basis of the theory.

Yet another limitation on predictions about specific technologies arises from the tendency of technologies to function in conjunction with other technologies, rather than as individual units. The Chinese invention of the printing press, for example, failed to result in a mass-print culture because their ideographic form of writing was incompatible with interchangeable type faces. Similarly, three-dimensional holography cannot replace two-dimensional television as the theory predicts until a satisfactory distribution system is developed.

Notwithstanding these limitations, however, the anthropotropic theory still seems on firm ground in its general thesis that media have developed, and will continue to develop, a closer correspondence to human or pre-technological modes of communication. The inability to precisely or completely define the construct of "pre-technological" environment means only that the construct requires more thought and investigation, and not that the construct is invalid or unusable. The evolution of media in response to elements of the pre-technological environment which have been clearly identified, such as color perception, amply demonstrate the viability of this construct and the theory that is built around it. Similarly, the intrusion of non-technological factors into the evolution of technology seems to pertain more to the time that the theory's predictions will take to become realized, than to the nature of the predictions themselves. The invention of video-phone is an inevitable prediction of anthropotropic theory; the mass use of video-phone in society is conditioned by economic and other cultural forces as discussed above.

* * *

The question of a pattern's existence is, of course, entirely separate from the question of whether the pattern's existence is socially desirable. Acceptance of the theory presented here as an essentially accurate description of events could thus serve as grounds for encouraging the further development of these events, or as grounds for mobilizing opposition to these developments. In other words, whether the evolution of media towards fuller replication of human patterns of communication is good or bad for humanity has nothing whatsoever to do with a judgement as to whether such an evolution is in fact occurring.

But the value of anthroptropic evolution does have relevance, if not to the establishment of the theory presented here, to the happiness of all human beings who must now inevitably live in some sort of relationship with media. Thus the social consequence of anthropotropic evolution deserves at least the brief consideration to be accorded it in the next and final chapter.

EPILOG

CHAPTER 9

THE CENTER BEHELD!

Jacques Ellul, that implacable critic of technology whose views have often been used here to contrast the views of the present study, would no doubt have to agree with much of the literal evidence that has been presented here. Ellul, after all, couldn't deny that more than 75 percent of American homes now have color televisions, or that the telephone has largely replaced the telegraph, or that stereophonic recordings have all but eliminated monaural recordings; nor could he deny the emphasis on "life" and "reality" in ads for new technologies ranging from the first phonographs to wall-size television screens, or the original descriptions of early photography and the telegraph, that speak of recapturing the world and reuniting the human race. But nor would Ellul wish to deny these very real events — for he would no doubt see them through different eyes, through eyes that confirmed his worst worries about technology.

In effect, Ellul's pessimism about technology would probably see at least two distinct and serious threats to humanity in the anthropotropic evolution of media — the first, as it turns out, more easily dismissed than the second.

APPROPRIATE MONSTERS

The first position that a critic of technology might take with regard to anthropotropic media has already been briefly alluded to in the first chapter: as technologies retrieve more and more of the real world of humanity, they become more and more of a threat to humanity, because they provide an ever more tempting — i.e., life-like, but yet non-life — alternative to human reality. [1] From this perspective, anthropotropic evolution becomes not a bettering but a worsening of technologies, a process in which primitive technologies, at least recognizable in their artificiality (and hence easier to deal with), are gradually replaced by technologies whose artificialities are much subtler and hence harder to withstand. The Frankenstein-monster metaphor has often been used to characterize Ellul's view of technology, but a Dracula motif would probably more closely fit his view of anthropotropic evolution: for nothing is so inimical to the living as the living-dead (life-like technology), which, walking among us, poses much more of a danger — in terms of conversion — than just the plain old dead (the obviously artificial, or primitive, technology). Moreover, since we are the creators of this living but non-living technology, the stake that would put an end to it would have to be driven through our hearts.

1. Sontag's view that photography is dangerous precisely because it is so close to reality, and hence offers a too comfortable substitute for reality,

is a recent example of such a criticism. See Ch. 5, n. 21 above.

But such a view of technology is both teleological and arbitrary at its core, for it assumes that, for some diabolic reason, technology will always stop short of totally retrieving all aspects of pre-technological, living environments. But nothing in the process of evolution herein described indicates that such an absolute barrier should exist somewhere down the road: indeed, the initial capturing of the visual world through the photograph, and the first instantaneous transmission of information over distances by the telegraph, seem to have bridged chasms far more unbridgable than those that still separate the instantaneous transmission of living bodies from the instantaneous transmission of the reflections and echoes of these bodies that has already been achieved. And without such a barrier to technological evolution — a barrier that in no way follows from the pattern described in the present study, and thus must be arbitrarily imposed from outside the system — anthropotropic technology at once is removed from the role of the friend who gets closer and closer to us, only to stab us in the back in the end. Instead, anthropotropic evolution becomes grounds for rejoicing over a genuine reconciliation between life and technology that has been long in coming.

THE DREAM AND THE ACT: THE IRREDUCIBLE DIFFERENCE

But perhaps such rejoicing might nonetheless be a bit premature. For even if (or when) technology were to retrieve all the forms and processes of living communication, there would still be an irreducible difference, as perceptive critics of technology would notice, between original pre-technological existence and the pre-technological existence retrieved by technology. The difference would not be between life and non-life, as Ellul might have argued above, but between life and extended life: for it has always been the very raison d'etre of technology to extend human power beyond the biological boundaries of the pre-technological world. Technology has evolved, as has been seen, from extending pre-technological reality by distorting it to extending pre-technological reality while conserving it — but the extending itself, whether distortive or replicative of the real world, is and always has been not a part of the actual pre-technological environment. Or, more precisely, extensions such as instantaneous speed and perfect recall have always existed in the imagination of the pre-technological world — but their actualization through technology has opened up the world of difference, as someone once put it, between the dream and the act.

Anthropotropic evolution may thus be defined as the development of unnatural extensions of the natural; and yet it has only been the technological transgression on the nature of space — the speed of communication — and not the transgression on time, or permanency, that has so worried the critics of technology. Thus, Lewis Mumford, who was the first to recognize the anthropotropic pattern in 1934, views the "maintenance of distance" as itself a necessary constituent of civilization, and fears the abolition of distance through transportation and communication technology as creating a psychological climate akin to "brain injury" (see ends of Chapters 3 and 5, above); McLuhan suggests that traveling at the speed of light deprives humans of their physical bodies, and thus suspends humans from all "natural" laws relating to their bodies, such as thou shalt not kill, etc. (end of Chapter 2 above; McLuhan here perhaps would not agree with the suggestion in Chapter 7 above that the ultimate mergence of transportation with communication will transmit humans at the speed of light not minus their bodies, but in their bodies); and Arnheim, recognizing that the danger (if it exists) lies not only in the speed but the fullness of communication, worries that "the more perfect our means of direct experience," the more endangered is our ability to reflect and think (Ch. 5, n. 38 above). The problem has perhaps been summed up in a recent essay by New York University professor Christine Nystrom on the significance of "transitional space," which suggests that the time we spend and the spaces we occupy in getting to where we're going teem with unnoticed activities[2] — activities which, it might be inferred, are crucial to human well-being, but which would of course have no place to occur when technology vanquishes distances, or gets us to where we're going instantly.

2. "Waiting: The Semantics of Transitional Space," *et cetera* 35 (September 1978): forthcoming.

A partial response to this criticism of technology would be to point out that many of the "distances" and delays in communication that advanced technology removes are themselves the products of more primitive technologies (such as writing and print) and not at all fundamentally human: thus, until the invention of writing, the pattern of delay in communication across distances was not common, simply because communication across any appreciable distance itself was not common; instead, humans communicated in the fluid, non-delayed environment of their immediate physical surroundings. Nonetheless, the transfer of this fluid immediacy to the world at large — where every human wish is but a command realizable through technology — may indeed unsettle some profound human foundation. Thus, one can sympathize with the concern of Mumford, McLuhan, et al. to the extent of admitting that no one can know, with certainty or even assurance, what the complete actualization of our dreams — specifically directed towards the banishment of distances — can have in store for us.

THE HUMAN OPTION

But should the consequences of such actualization prove undesirable, they would still be greatly mitigated by the human ability to control or exercise options upon technology, a human power which has been demonstrated to a greater or lesser degree in many technologies throughout history. An excellent example of this human ability to control undesirable consequences of technology is provided by the window, a medium Ed Wachtel has recentlycalled attention to in an essay that explores the window's influence upon Western ways of seeing and art.[3] Wachtel's thesis is that the window, as an early organizer of visual perceptions and a device which overcame the "blockage" of the wall, functions as the "archetype" for subsequent media such as photography, television, etc., which also organize visual perceptions and overcome blockages of time and space. Archetypical in this way or not, the window certainly seems typically anthropotropic: it replaces the more primitive wall (which extends human skin at the expense of blocking perception of the outside), and continues the extensional function of the wall while retrieving the lost pre-technological element of the "outside." The window thus offers the protection of the wall while retrieving at least part of the pre-wall perception.

3. "The Influence of the Window On Western Art and Vision," *The Structurist* 17/18 (October 1978): forthcoming.

But what is of greater concern to the present discussion (and unaddressed by Wachtel) are the problems the window creates for its users, and how we have handled these problems. For in inventing the window, we also invented the Peeping Tom — the unintended consequence of a technology that allows us to see through walls is that people can look in at us. And what was the human response to this human-created problem? Did we meekly surrender our privacy to the irresistible seductress/tyrant of *La Technique*? Should we have smashed the window in its infancy before it utterly disrupted our sexual mores?

Instead, we invented the window shade, the curtain, and all manner of easily usable window coverings. That is, we developed auxiliary technologies which allowed us to control the worst consequences of the window technology, without unduly sacrificing its benefits. In effect, we gave ourselves an option on the use of the window.

And we have been exercising similar options with a variety of technologies for some time now. Those who wish to leave Ellul's "lunar world of stone" may live in the country; those who dislike autos are free to walk or bicycle if they choose to live close to where they're going; those that dislike working at a desk may choose any number of jobs that give them access to the outdoors; and so forth. While it is true that economic wherewithal often severely limits these options, the point is that technological consequences are not, in principle, inescapable — we do have the option, more times than not, of at least partially cutting ourselves loose if we so desire.

Moreover, communication technology has traditionally been even more controllable than other technologies. There are numerous people today who choose to read rather than watch television, or write letters rather than talk on the phone. The ringing phone itself — often depicted as the medium you can't say no to — is actually no more requiring of attention than the personal knock on the door (perhaps less so, for the phone can at least be taken off the hook). And the videophone, as suggested earlier, need not guarantee unwanted eyes access to your bedroom — for such devices would surely leave you the option of turning your camera off.

And so it should be with the almost god-like powers that communication/transportation technologies may soon invest us with. We'll turn them off when we don't want them, and probably be able to modify them if they hurt us. The ability to travel from New York to London instantly should no more prevent us from walking around the block if we so desire, than the ability to travel from New York to London in three and a half hours now prevents us from walking around the block. Being "in touch" night and day should pose no problem to us if we have the ability to cut ourselves off when we wish — a strategy well understood by people who have learned to take their telephones off the hook during inopportune moments. The exercise of such options is made possible by a characteristic of media evolution that has perhaps been insufficiently emphasized in the present work: namely, that media evolution tends to replace rather than obliterate primitive technologies (e.g., courses in reading and writing hieroglyphics are available in many universities), and supplement rather than completely substitute for pre-technological communication patterns. Just as the biological world embraces all complexities of organisms, so the world of evolving technologies allows for the use of almost any earlier technology, or any non-technological mode of communication, should it be humanly desired.

Still, as the power becomes more immense, the margin of error reduces. Hall's example of the atomic bomb as a dysfunctional extension is frightening with due cause — because technologies have indeed invested us with the awful power to utterly destroy ourselves. And yet, each individual throughout history has always had the power to totally destroy him or herself, and has rarely blundered into using it. This suggests that just as the percentages for life may be with us on an individual basis (i.e., most people do not commit suicide), they may be with us on a collective basis. In effect, the survival of our species to date is the single best, though certainly not infallible, argument for the survival of our species in the future.

The worst consequences of anthropotropic media, then, are both unspecified and probably controllable. And how do these dangers measure up to the benefits? The thrust of everything in the present study — not to mention the framework of even the severest critics of technology like Ellul — shows that technology was not something that was artificially imposed upon us from the outside. Rather, its various manifestations were dreamed about, imagined, fervently desired, long before they were actually created. And why should it have not been so? What is inhuman about hearing the voice, seeing the face, or reading the thoughts of a loved one when they would not otherwise be physically present? How would our humanity be lessened by seeing a three-dimensional hologram of Socrates speaking instead of reading Plato's description of what he said? In fact, what could be more human — and better for humanity — then to extend our human life anyplace our minds imagine it could be?

The great media explorers whose works have served as the basis for this study have understood this — otherwise they would not be so consistent in their at least partial recognition of an anthropotropic pattern on much less evidence than exists in 1978 — but like prophets first blinded by the signs of God they have been unable to fully accept it. And yet glints of recognition inevitably peek through the nervous rejections. And so it is fitting that Lewis

Mumford, whose views have so often served to oppose the implications of anthropotropic evolution discerned here, but who nonetheless was the first to notice such an emergence more than 40 years ago, should have the last word in this matter. He writes of the cosmic black hole as a metaphor for his near despairing fears about technology and humanity, but then concludes that

> . . . perhaps, with a further twist of the ring, the impenetrable Black Hole may prove a shadow of a brighter sun. Even the notion of an Explosion and an Implosion, a "beginning" and "ending," may be only a very human metaphor, which the universe, for reasons of its own, neither recognizes or exhibits. On that ultimate skepticism my own faith blithely flourishes. Let the curtain rise on the twenty-first century — *and After!* [Italics in original.][4]

4. "Reflections: Prologue to our Times," *New Yorker*, 10 March 1975, p. 63.

Indeed, let it rise!

BIBLIOGRAPHY

BOOKS

Abercrombie, John. *Inquiries Concerning the Intellectual Powers, and the Investigation of Truth.* Original ed., n.p., 1833; reprint ed. Boston: Otis, Broaders, 1843.

Agassi, Joseph. *The Continuing Revolution.* New York: McGraw-Hill, 1968.

Allport, Gordon. *Personality.* New York: Holt, 1937.

Appleton, Victor. *Tom Swift and his Photo Telephone.* New York: Grosset & Dunlap, 1914.

Arnheim, Rudolf. *Film as Art.* Berkeley and Los Angeles: University of California Press, 1957.

Asimov, Isaac. *Foundation.* New York: Avon, 1951.

Barnouw, Erik. *The Tube of Plenty.* New York: Oxford University Press, 1975.

Baumer, Franklin L. *Modern European Thought.* New York: Macmillan, 1977.

Bazin, André. *What is Cinema?* Foreword by Jean Renoir. Compiled and translated by Hugh Gray. Original ed., n.p., 1958-1965; trans. ed., Berkeley and Los Angeles: University of California Press, 1967.

Beals, Ralph J., and Hoijer, Harry. *An Introduction to Anthropology.* New York: Macmillan, 1965.

Birx, James H. *Pierre Teilhard de Chardin's Philosophy of Evolution.* Springfield, Illinois: Thomas, 1972.

Butler, Samuel. *Erewhon*. Original ed., n.p., 1872; reprint ed., London: Dent, 1965.

_____. *Life and Habit*. Original ed., n.p., 1878; reprint ed., New York: Dutton, 1910.

Chomsky, Noam. *Language and Mind*. 2nd ed. Original ed., 1968; rev. ed., New York: Harcourt Brace Jovanovich, 1972.

_____. *Reflections On Language*. New York: Pantheon, 1975.

Christensen, Erwin O. *The History of Western Art*. New York: Mentor, 1959.

Coleridge, Samuel Taylor. *Biographia Literaria. Vol. 2*. Edited by J. Shawcross. Original ed., n.p., 1817; reprint ed., London: Oxford University Press, 1907

Dawkins, Richard. *The Selfish Gene*. New York: Oxford University Press, 1976.

DeGeorge, Richard, and DeGeorge, Fernande, eds. *The Structuralists from Marx to Levi-Strauss*. New York: Anchor, 1972.

deHaan, David. *Antique Household Gadgets and Appliances*. Woodbury, N.Y.: Barron's, 1977.

Derry, T.K., and Williams, Trevor I. *A Short History of Technology*. New York: Oxford University Press, 1961.

Eisenstein, Sergei. *Notes of a Film Director*. Note by Richard Griffith. Compiled and edited by R. Yurenev. Translated by X. Danko. Original ed., n.d. [shortly after 1948]; new ed., New York: Dover, 1970.

Ellul, Jacques. *The Technological Society*. Translated by John Wilkinson. Original ed., n.p., 1954; trans. ed., New York: Knopf, 1964.

Findlay, J. N. *Hegel: A Re-Examination*. London: George Allen &

Unwin, 1958.

Ford, Kenneth W. *Basic Physics*. Waltham, MA: Blaisdeli, 1968.

Freud, Sigmund. *Civilization and Its Discontents*. New ed. Translated and edited by James Strachey. Original ed., n.p., 1930; new e d., New York: Norton, 1961.

Fuller, R. Buckminster. *Nine Chains to the Moon*. Carbondale, Illinois: Southern Illinois University Press, 1938.

Gardner, Howard. *The Quest for Mind*. New York: Knopf, 1973.

Gombrich, E.H. *Art and Illusion*. Rev. 2nd ed. Princeton, N.J.: Princeton University Press, 1972.

Gordon, Cyrus. *Before Columbus*. New York: Crown, 1971.

Gordon, George. *Persuasion*. New York: Hastings House, 1971.

Gregor, Arthur. *Bell Laboratories*. New York: Charles Scribner's, 1972.

Hall, Edward T. *Beyond Culture*. New York: Anchor, 1976.

_____. *The Silent Language*. New York: Fawcett, 1959.

Harmon, Gilbert, ed. *On Noam Chomsky*. New York: Anchor, 1974.

Head, Sidney W. *Broadcasting in America*. 3rd ed. Boston: Houghton Mifflin, 1976.

Herm, Gerhard. *The Phoenicians*. Translated by Caroline Hillier. New York: William Morrow, 1975.

Hiebert, Ray Eldon; Ungurait, Donald F.; and Bohn, Thomas. *Mass Media*. New York: David McKay, 1974.

Hilgard, Ernest. *Introduction to Psychology*. 3rd ed. New York: Harcourt, Brace & World, 1962.

Innis, Harold A. *The Bias of Communication*. Introduction by Marshall McLuhan. Original ed., n.p., 1951; reprint ed., Toronto: University of Toronto Press, 1964.

_____ . *Empire and Communications*. Revised by Mary Q. Innis. Foreword by Marshall McLuhan. Original ed., London: Oxford University Press, 1950; rev. ed., Toronto: University of Toronto Press, 1972.

Jaynes, Julian. *The Origin of Consciousness in the Breakdown of the Bicameral Mind*. Boston: Houghton Mifflin, 1976.

Jensen, Hans. *Sign, Symbol, and Script*. 3rd. ed. Revised and enlarged. Translated by George Unwin. New York: Putnam's. 1969.

Josephson, Matthew. *Edison*. New York: McGraw-Hill, 1959.

Koestler, Arthur. *The Act of Creation*. New York: Macmillan, 1964.

_____ . *Janus: A Summing Up* . New York: Random House, 1978.

_____ . *The Sleepwalkers*. New York: Grosset & Dunlap, 1959.

Kracauer, Siegfried. *Theory of Film: The Redemption of Physical Reality*. New York: Oxford University Press, 1960.

Kuhn, Thomas S. *The Structure of Scientific Revolutions*. 2nd ed. Chicago: University of Chicago Press, 1970.

Kuhns, William. *The Post-Industrial Prophets*. New York: Harper Colophon, 1971.

Langer, Susanne. *Feeling and Form*. New York: Charles Scribner's, 1953.

Leach, Edmund. *Claude Levi-Strauss*. Rev. ed. New York: Viking, 1974.

Lorenz, Konrad. *On Aggression*. New York: Bantam, 1963.

McLellan, David. *Karl Marx*. New York: Viking, 1975.

McLuhan, Marshall. *The Gutenberg Galaxy*. New York: Mentor, 1962.

──────. *Understanding Media: The Extensions of Man*. 2nd. ed. New York: Mentor, 1964.

McLuhan, Marshall; Hutchon, Kathryn; and McLuhan, Eric. *City As Classroom*. Agincourt, Ontario: Society of Canada, 1977.

Maddox, Brenda. *Beyond Babel*. New York: Simon & Schuster, 1972.

Magee, Bryan, *Karl Popper*. New York: Viking, 1973.

Marshack, Alexander. *The Roots of Civilization*. New York: McGraw-Hill, 1972.

Marx, Karl. *Capital*. Edited by Frederick Engels. 4th ed. Edited, revised, and amplified by Ernest Unterman. Original English ed., n.p., 1867; rev. ed., New York: Modern, 1906.

Mast, Gerald. *A Short History of the Movies*. New York: Pegasus, 1971.

Mast, Gerald, and Cohen, Marshall. *Film Theory and Criticism*. New York: Oxford University Press, 1974.

Miller, Jonathan. *Marshall McLuhan*. New York: Viking, 1971.

Monaco, James. *How to Read a Film*. New York: Oxford University Press, 1977.

Mumford. Lewis. *The City in History*. New York: Harcourt Brace Jovanovich, 1961,

──────. *Interpretations and Forecasts*. New York: Harcourt Brace Jovanovich, 1973.

———. *The Myth of the Machine. Vol. 1: Technics and Human Development*. New York: Harcourt Brace Jovanovich, 1966.

———. *The Myth of the Machine. Vol. 2: The Pentagon of Power*. New York: Harcourt Brace Jovanovich, 1970.

———. *Technics and Civilization*. New York: Harcourt Brace, 1934.

Newhall, Beaumont. *The Daguerreotype in America*. N.p.: Duell, Sloan & Pearce, 1961.

Ong, Walter. *Interfaces of the Word*. Ithaca, N.Y.: Cornell University Press, 1977.

Patrouch, Joseph F. *The Science Fiction of Isaac Asimov*. New York: Doubleday, 1974.

Popper, Karl R. *Objective Knowledge*. London: Oxford University Press, 1972.

Popper, Karl R., and Eccles, John C. *The Self and Its Brain*. New York: Springer International, 1977.

Rose, H. J. *A Handbook of Greek Mythology*. New York: Dutton, 1959.

Sagan, Carl. *The Cosmic Connection*. New York: Dell, 1973.

———. *The Dragons of Eden*. New York: Random House, 1977.

Sontag, Susan. *On Photography*. New York: Farrar, Straus, and Giroux, 1977.

Spencer, Herbert. *First Principles*. 4th. ed. Originated., n.p., 1864; rev. ed., New York: Appelton, 1896.

Stroke, George W. *An Introduction to Coherent Optics and Holography*. 2nd ed. New York: Academic Press, 1969.

Taischoff, Sol, ed. *Broadcasting Yearbook 1977*. Washington,

D.C.: Lawrence B. Taischoff, 1977.

_____ , ed. *Broadcasting Yearbook 1978*. Washington, D.C.: Lawrence B. Taischoff, 1978.

Teilhard de Chardin, Pierre. *The Future of Man*. Translated by Norman Denny. Original ed., n.p., 1959; trans. ed., New York: Harper & Row, 1964.

_____ . *The Phenomenon of Man*. Translated by Bernard Wall. Introduction by Julian Huxley. Original ed., n.p., 1955; trans. ed., New York: Harper & Row, 1959.

Tudor, Andrew. *Theories of Film*. New York: Viking, 1973.

Wiener, Norbert. *Cybernetics*. 2nd ed. New York: M.I.T. Press and John Wiley, 1961.

Wilson, Edward O. *Sociobiology*. Cambridge, MA: Harvard University Press, Belknap Press, 1975.

ARTICLES AND MISCELLANEOUS PRINTED MATERIAL

A.C. Nielson Company. *Nielson Television 1977*. Northbrook, IL.: A.C. Nielson, 1977.

Advertisement (direct mail) for the *CoEvolution Quarterly*, Sausalito, California, 1976.

Advertisement (magazine) for Advent Video Beam. *The New York Times Magazine*, 4 December 1977, p. 155.

Advertisement (magazine) for Columbia Phonograph Company. "The Gramaphone Grand." New York, 1899. (Periodical unidentified.)

Advertisement (magazine) for Columbia Records. "Columbia Grafonola." *The Delineator*, October 1917, p. 41.

Advertisement (periodical) for Bell System. "What's next, Professor Bell?" *New York Review of Books,* 28, October 1976, p. 8.

Baur, Stuart. "Kneedeep in the Cosmic Overwhelm with Carl Sagan." *New York*, 1 September 1975, pp. 26, 28-32.

Blechman, Robert. *Myth as Advertising: A Structural Analysis of Selected American Television Advertisements Using a Methodology Based on the Theories of Claude Levi-Strauss*. Ph.D. dissertation, New York University, 1978.

Boyle, W.S. "Light-Wave Communications." *Scientific American* 237 (August 1977): 40-48.

Brown, Roger. "How Shall a Thing Be Called?" In *Readings in Child Development and Personality*, pp. 267- 276. Edited by Paul H. Mussen, John J. Conger, and Jerome Kagan. New York: Harper & Row, 1965.

Browne, Malcolm W., and Winfrey, Carey. "*The Times* Enters a New Era of Electronic Printing." *The New York Times*, 3 July 1978, pp. 21, 38.

Buber, Martin. "Between Man and Man: The Realms." In *The Human Dialogue*, pp. 113-117. Edited by Floyd W. Matson and Ashley Montagu. New York: Free Press, 1962.

Butler, Samuel. "Darwin Among the Machines." *Christchurch Press*, New Zealand, 13 June 1863. Reprinted in The Note-Books of Samuel. Butler, pp. 35- 40. Edited by Henry Festing Jones. New York: AMS, 1968.

_____. "Lucubratio Ebria." *Christchurch Press,* New Zealand 29 July 1865. Reprinted in *The Note-Books of Samuel Butler*, pp. 40-46. Edited by Henry Festing Jones. New York: AMS, 1968.

Carpenter, Edmund. "The New Languages." In *Explorations in Communication*, pp. 162-179. Edited by Marshall McLuhan and Edmund Carpenter. Boston: Beacon, 1960.

Clarke, Arthur C. "Communications in the Second Century of the Telephone." In *The Telephone's First Century — and Beyond*, pp. 83-112. Preface and Afterword by John D. deButts. Introduction by Thomas E. Bolger. New York: Crowell, 1977.

Consumer Reports Buying Guide Issue 1978. Mt. Vernon, N.Y.: Consumers Union, 1977.

Dessy, Raymond E. "Microprocessors? — An End User's View." In *Electronics: The Continuing Revolution*, pp.

138-145. Edited by Philip H. Abelson and Allen L. Hammond. Washington, D.C.: American Association for the Advancement of Science, 1977.

Dipboye, Marilyn. "Wallace Nutting, Photographer." *American Collector*, November 1977, pp. 27-30.

Ditlea, Steve. "What is a Hologram? How Does It Float in Mid Air... And Is It An Art?" *Ms.*, December 1976, pp. 34-39.

Gould, Stephen Jay. "Koestler's Solution." Review of *Janus: A Summing Up*, by Arthur Koestler. *New York Review of Books*, 20 April 1978, pp. 35-37.

Gumpert, Gary. "The Rise of the Mini-Comm." *Journal of Communication* 20 (September 1970): 280-290.

Haratonik, Peter. *Toward the Biotechnic Order: A Study of the Writings of Lewis Mumford*. Ph.D. dissertation, New York University, forthcoming.

Hawkes, Nigel. "British May Use Telephones, TVs, to Tap Data Bank." *Science*, 9 July 1978, pp. 33-34.

Heilbroner, Robert L. "Inescapable Marx." *New York Review of Books*, 29 June 1978, pp. 33-37.

Hogarth, S. H. "Three Great Mistakes." *Blue Bell*, November 1926.

"In the FCC's Lap: FM Quad, AM Stereo, TV Stereo." *Broadcast Management/Engineering*, October 1977, p. 53.

Isaac, Glynn. "The Food-sharing Behavior of Protohuman Hominids." *Scientific American* 238 (April 1978): 90-108.

Jaynes, Gregory. "A New TV Idea: Shows for Old People." *The New York Times*, 7 July 1978, p. A7.

Kolm, Henry. "An Electromagnetic 'Slingshot' for Space Propulsion." *Technology Review* 79 (June 1977): 60-66.

Korzenny, Felipe. "A Theory of Electronic Propinquity." *Communications Research* 5 (January 1978):3-23.

Lachenbruch, David. "The New Boom in Video-cassette Recorders." *House and Garden*, November 1977, pp. 64-65.

Levinson, Paul. "'Hot' and 'Cool' Redefined for Interactive Media." *Media Ecology Review* 4 (February 1976): 9-11.

_____ . Review of *Reflections On Language*, by Noam Chomsky. *Media Ecology Review* 4 (May 1976): 24-26.

_____ . "Toy, Mirror, and Art: The Metamorphosis of Technological Culture." *et cetera* 34 (June 1977) : 151-167.

Lieberman, Ben. Untitled essay in *McLuhan: Hot and Cool*, pp. 222-225. Edited by Gerald Emanuel Stearn. New York: Dial, 1967.

Macdonald, Dwight. Untitled essay in *McLuhan: Hot and Cool*, pp. 203-211. Edited by Gerald Emanuel Stearn. New York: Dial, 1967.

McLuhan, Marshall. Comments in *McLuhan: Hot and Cool*, p. xiii. Edited by Gerald Emanuel Stearn. New York: Dial, 1967.

_____ . "The Hemispheres and the Media." Centre for Culture and Technology, University of Toronto, Toronto, Canada, 12 February 1977. (Typewritten.)

_____ . Introduction to *The Bias of Communication* by Harold A. Innis. Reprint ed., Toronto: University of Toronto Press, 1964.

_____ . "Laws of the Media." Introduction by Paul Levinson, *et cetera* 34 (June 1S77): 173-179.

_____ . "Note on Discarnate Man." Centre for Culture and Technology, University of Toronto, Toronto, Canada, 10 May 1977. (Typewritten.)

_____. Review of *The Origin of Consciousness In the Breakdown of the Bicameral Mind* by Julian Jaynes. Centre for Culture and Technology, University of Toronto, Toronto, Canada, 3 June 1977. (Typewritten.)

_____. Review of *Television: Technology and Cultural Form* by Raymond Williams. *Technology and Culture* 19 (April 1978): 259-261.

McLuhan, Marshall, and Logan, R.K. "Alphabet, Mother of Invention." *et cetera* 34 (December 1977): 373-383.

Maeroff, Gene I. "Reading Achievement of Children in Indiana Found as Good as in '44." *The New York Times*, 15 April 1978, p. 10.

Mason, Roy. "Biological Architecture." *The Futurist* 11 (June 1977): 140-147.

Mead, George Herbert. "Thought, Communication, and the Significant Symbol." In *The Human Dialogue*, pp. 397-403. Edited by Floyd W. Matson and Ashley Montagu. New York: Free Press, 1967.

Medawar, Peter. "What's Human About Man Is His Technology." *Smithsonian* 4 (May 1973): 22-28.

Meyrowitz, Joshua. "Television and Interpersonal Behavior: Codes of Perception and Response." In *Inter/Media: Interpersonal Communications in a Media World*. Edited by Robert Cathcart and Gary Gumpert. New York: Oxford University Press, forthcoming.

Milgram, Stanley. "The Image-Freezing Machine." *Psychology Today* 10 (January 1977): 50-45 ff.

Miller, Stewart E. "Photons in Fiber for Telecommunication." In *Electronics: The Continuing Revolution*, pp. 167-172. Edited by Philip H. Abelson and Allen L. Hammond. Washington,

D.C.: American Association for the Advancement of Science, 1977.

Morriss, James. "Comments and Reflections on a Theory of Linguistic Consciousness." *et cetera* 35 (September 1978): forthcoming.

Mumford, Lewis. "Reflections: Prologue to our Times." *The New Yorker*, 10 March 1975, pp. 42-63.

"NBC to Discontinue Next Year Its All-News Service for Radio." *The New York Times*, 4 November 1976, p. 78.

New World Dictionary/Second Collegiate edition, 1974. S.v. "television".

Nystrom, Christine L. *Toward A Science of Media Ecology*. Ph.D. dissertation, New York University, 1974.

_____ . "Waiting: The Semantics of Transitional Space." *et cetera* 35 (September 1978): forthcoming.

NYU - Reading Consortium. "Final Report: Berks Community Television." New York: NYU-Reading Consortium, forthcoming.

Potter, Robert J. "Electronic Mail." In *Electronics: The Continuing Revolution*, pp. 90-94. Edited by Philip H. Abelson and Allen L. Hammond. Washington, D.C.: American Association for the Advancement of Science, 1977.

Products That Think, Northbrook, Illinois: J S & A National Sales Group, 1977-1978.

Reif, Rita. "Collectors Focus on Daguerreotypes." *The New York Times*, 9 October 1977, sec. 2, p. 34.

Rensberger, Boyce. "Roots of Writing Traced Back More Than 10,000 Years." *The New York Times*, 9 July 1977, pp. 19, 20.

_____ . "The World's Oldest Works of Art." *The New York*

Times Magazine, 21 May 1978, pp. 26-29, ff.

Sartre, Jean-Paul. "The Work of Art." In *Aesthetics*, pp. 32-38. Edited by Harold Osborne. London: Oxford University Press, 1972.

Schmeck, Harold M., Jr. "Scientists Talk on Solar Power Stations Aloft and a Super Subway." *The New York Times*, 14 February 1978, p. 12.

Searle, John. "Chomsky's Revolution in Linguistics." In *On Noam Chomsky*, pp. 2-33. Edited by Gilbert Harmon. New York: Anchor, 1974.

"A Space-Age Treadmill." *Forbes*, 1 August 1977, pp. 59- 60.

Stent, Gunther. "Limits to the Scientific Understanding of Man." *Science*, 21 March 1975, pp. 1052-1057.

Sullivan, Walter. "How to Search for Life on Mars." *The New York Times*, 1 August 1976, sec. 4, p. 8.

_____ "Tests to Seek Life on Mars Begin." *The New York Times*, 25 July 1976, pp. 1, 28.

Taylor, Richard. "Towards Communications Policies for the 21st Century." *Communications Tomorrow*, September 1977, pp. 1, 3.

Teltsch, Kathleen. "Mondale, at U.N., Offers Nations Equipment to Help Avert Conflicts." *The New York Times*, 25 May 1978, pp. Al, A16.

Thaler, Pat, and Shapiro, Sonya. "New Routes to a College Degree." *New York*, 29 August 1977, p. 40.

Thorpe, W. H. "Arthur Koestler and Biological Thought." In *Astride the Two Cultures*, pp. 50-68. Edited by Harold Harris. New York: Random House, 1976.

Wachtel, Ed. "The Influence of the Window on Western

Art and Vision." *The Structurist* 17/18 (October 1978): forthcoming.

_____ . *The Transformation of Visual Space: A Theory of the Relationship Among Technology, Space Conception, and Culture Change*. Ph.D. dissertation, New York University, forthcoming.

Weaver, Warren. "The Mathematics of Communication." In *Communication and Culture*, pp. 13-24. Edited by Alfred G. Smith. New York: Holt, Rinehart, and Winston, 1966.

Zilkha, Michael. "Future Hock." *New York*, 19 April 1976, pp. 66-68.

NON-BOOK, NON-PERIODICAL MEDIA

Advertisement (billboard) for Panasonic. "Quintrix II . . . So lifelike you'll feel you're part of the picture." West Side Highway, New York City, 1977-1978.

Advertisement (televised) for Memorex Tape. "Is it live, or is it Memorex?" WNBC-TV; 1977-1978.

Advertisement (televised) for Pioneer Systems. "If man had a Pioneer cassette . . . " WNBC-TV, 23 April 1977, 12:35 AM.

Advertisement (televised) for Pioneer Systems. "We bring it back alive." WNBC-TV, 9 October 1977, 12:14 AM.

Advertisement (televised) for RCA. "Colortrak." WCBS-TV, 1977-1978.

Advertisement (televised) for Scotch Brand Recording Tape. "The truth comes out." WNBC-TV, 1977-1978.

Advertisement (televised) for theatrical production of *Grease*. WNBC-TV, Summer 1978.

Dracula. Broadway play produced at Martin Beck Theater, New York City, 1978.

"Eyewitness News." WABC-TV, 13 February 1978, 6:00-7:00 PM.

Field, Frank. "News Center Four." WNBC-TV, 1 November 1976, 6:25 PM.

Franks, Michael. "Night Moves." Song on The Art of Tea (LP record). Burbank, California: Warner Bros. Records, 1976.

Gross, Talia. WNET-TV, New York City. Telephone Interview, 2 May 1978.

Klein, Paul. The New School for Social Research, New York City. Lecture, Fall 1975.

Murray, Bill. "Saturday Night Live." WNBC-TV, 2 April 1978, 12:04 AM.

Potts, Robert. "News Center Four." WNBC-TV, 3 March 1977, 6:50 PM.

Simon, Paul. "Kodachrome." Song on *There Goes Rhymin' Simon* (LP record). New York: Columbia Records, 1973.

GLOSSARY

TERMS USED IN THIS DISSERTATION

abstraction: a representation that bears only an arbitrary connection to the object represented; e.g., the word "tree" is an abstraction of the object tree in that the word has no literal connection to the object

access: the availability of information in a communication system to any user of that system

anthropotropic: an adjective introduced in the present study to describe the evolution of media towards greater replication of human communication patterns ("anthropo" = human; "tropic" = towards) (see "Stage C" below)

communication: the transmission of information; entails both the "process" of transmission, and the "content" transmitted (see "content" and "process" below)

content: the material or information that is transmitted in communication; content is "what" is transmitted

convergence: the tendency noted in this study of individual media to evolve into integrated units; e.g., television may be considered a convergence of radio and film in that television combines the electronic delivery system of radio with the audio-visual content of film

ecological niche: the attainment by a medium of

a close correspondence to some aspect of the human communication environment, thereby assuring the medium's likely survival; e.g., radio has attained an ecological niche in presenting sounds without sight, which corresponds to the human communication pattern of eavesdropping

evolution: a change of events or phenomena over time (in the case of the present study, media) in a systematic pattern amenable to observation and theorization

extension: the communication beyond immediate physical surroundings perceivable by eyes, ears, etc. (see "space extension" and "time extension" below)

falsification: a philosophical approach to knowledge utilized in the present study which suggests that scientific knowledge can accrue only by the elimination of falsities from the body of knowledge; the method of science thus becomes a search for all apparent contradictions to the hypothesis

interaction: a type of communication in which the sender and receiver of information readily interchange roles; e.g., a conversation in person or on the telephone (see "observation" below)

medium: any artificial device used for communication; the term is used in the present, study interchangeably with "technological communication"

mobility: the ability of any user of media to freely move anywhere in the environment without sacrificing access or availability of information; mobility is a function of the portability of media

model: phenomena organized into a logical, conceptual pattern; in the present study, the evolution of media is described by a three-stage model (see "Stage A," "Stage B," and "Stage C" below)

observation: a type of one-way communication in which the human only receives information without influencing the transmission; e.g., watching a sunset or watching a movie (see "interaction" above)

pre-technological communication: the communication or transmission of information occurring without technology or media; unextended communication; term used interchangeably in the present study with "natural communication," "pre-technological reality," "human communication," and "Stage A"

primitive technology: technology that extends beyond biological boundaries only by alteration or distortion of the information communicated; e.g., letters permit people to communicate across vast distances, but minus their voices, physical appearances; also referred to in present study as "Stage B" technology

process: the method by which communication takes place; "how" the information is transmitted (see "content" above)

space extension: the ability of technologies to rapidly transmit information across large distances: "speed" of communication (see "time extension" below)

Stage A: the first part of a three-stage model developed in the present study to describe the evolution of media; communication in Stage A is non-technological; information is transmitted solely by human senses, perceptions, etc., and does not extend beyond immediate physical surroundings or the range of human memory; all characteristics of the real world such as color, depth, etc., are present (see "pre-technological communication" above)

Stage B: the second stage in which technologies extend communication beyond immediate physical surroundings, but only by distorting the information or by leaving much of these physical surroundings out of the communication process; in effect, this is a "trade-off" stage in which extension beyond biological boundaries is paid for with a sacrifice of the real world; e.g., telegraph communicates instantly across vast distances, but without voices, images, etc. (see "primitive technology" above)

Stage C: the third and culminating stage of media evolution in which technologies continue the extension beyond biological boundaries of Stage B, but recapture many of the elements of real life that were lost in the transition from Stage A to Stage B; thus telephone continues the extension of telegraph but restores the voice to the communication process, as video-phone continues the extension of the telephone but restores the visual image (see "anthropotropic" above)

technological communication: the transmission of information through or with any artificial device; used interchangeably in the present study with "medium"

technology: products or artifacts created by humans for the purpose of controlling the environment; since the present study is concerned with communication technologies or media, the term "technology" is often used here interchangeably with "medium" and "technological communication"

theory: a series of interrelated concepts and definitions that present a systematic view of identified phenomena (the phenomena in this case being media), so as to describe and explain their past and present occurrence, and predict their occurrence in the future

time extension: the ability of technologies to preserve information across time; "permanency" of communication (see "space extension" above)

ABOUT THE AUTHOR

Paul Levinson, PhD, is Professor of Communication & Media Studies at Fordham University in NYC. His nonfiction books, including *The Soft Edge* (1997), *Digital McLuhan* (1999), *Realspace* (2003), *Cellphone* (2004), and *New New Media* (2009; 2nd edition, 2012), have been translated into fifteen languages. His science fiction novels include *The Silk Code* (winner of Locus Award for Best First Science Fiction Novel of 1999), *Borrowed Tides* (2001), *The Consciousness Plague* (2002), *The Pixel Eye* (2003), *The Plot To Save Socrates* (2006), *Unburning Alexandria* (2013), *Chronica* (2014), and *It's Real Life: An Alternate History of The Beatles* (2024). His novelette, "The Chronology Protection Case," was made into a short movie available on Amazon Prime. His alternate history short story about The Beatles, "It's Real Life," was made into a radio play, won The Mary Shelley Award for Outstanding Fiction, and was expanded into a novel *It's Real Life: An Alternate History*

of The Beatles published in 2024. His novelette, "Robinson Calculator," was published in the *Robots Through the Ages* anthology in July 2023. He was President of the Science Fiction Writers of America (SFWA) 1998-2001. He has appeared on CNN, MSNBC, Fox News, the Discovery Channel, National Geographic, the History Channel, NPR, and numerous TV and radio programs. His 1972 LP, *Twice Upon a Rhyme*, was reissued in 2010. His first LP since 1972, *Welcome Up: Songs of Space and Time*, was released in 2020 by Old Bear Records and Light in the Attic Records.

Copyright 1978, 2017 by Paul Levinson

ISBN 978-1-56178-061-7

Published by Connected Editions

THE FOLLOWING BOOKS BY PAUL LEVINSON AVAILABLE IN PRINT AND KINDLE:

Science fiction:

It's Real Life: An Alternate History of The Beatles

Loose Ends (time travel) series (complete):
Loose Ends, Little Differences, Late Lessons, Last Calls; or, all four in The Loose Ends Saga

Sierra Waters (time travel) series:
The Plot to Save Socrates, Unburning Alexandria, Chronica

Phil D'Amato forensic detective series:
The Chronology Protection Case, The Copyright Notice Case, The Silk Code, The Consciousness Plague, The Pixel Eye

Ian's Ions and Eons (three time travel novelettes)

Exo-Genetic Engineers series
The Orchard, The Suspended Fourth

Double Realities series
The Other Car, Extra Credit, The Wallet, The P&A

Borrowed Tides

The Kid in the Video Store

Peter Brown Called: Tales of SciFi and Music

Urban Corridors: Fables and Gables

Marilyn and Monet

Robinson Calculator

Nonfiction and Science Fiction

Touching the Face of the Cosmos: On the Intersection of Space Travel and Religion
an anthology of essays and science fiction stories, including the final interview with John Glenn, an essay by Guy Consolmagno, SJ (the "Pope's Astronomer"), and contributions from leading thinkers about the role of religion in space travel

Nonfiction:

The Soft Edge: A Natural History and Future of the Information Revolution

Digital McLuhan: A Guide to the Information Millennium

Realspace: The Fate of Physical Presence in the Digital Age, On and Off Planet

Cellphone: The Story of the World's Most Mobile Medium, and How It Has Transformed Everything! (hardcover)

New New Media

From Media Theory to Space Odyssey: Petar Jandrić interviews Paul Levinson

McLuhan in an Age of Social Media

Fake News in Real Context

Cyber War and Peace

Paul Levinson Interviews Rufus Sewell about The Man in the High Castle